Calculus for Computer Graphics

John Vince

Calculus for Computer Graphics

Second Edition

 Springer

John Vince
Bournemouth University
Hereford, UK

ISBN 978-3-030-11375-9   ISBN 978-3-030-11376-6   (eBook)
https://doi.org/10.1007/978-3-030-11376-6

Library of Congress Control Number: 2019930968

This Springer imprint is published by the registered company Springer Nature Switzerland AG
The registered company address is: Gewerbestrasse 11, 6330 Cham, Switzerland

This book is dedicated to my wife He$^i\partial i\pi$.

Preface

Calculus is one of those subjects that appears to have no boundaries, which is why some Calculus books are so large and heavy! So when I started writing the first edition of this book, I knew that it would not fall into this category. It would be around 200 pages long and take the reader on a gentle journey through the subject, without placing too many demands on their knowledge of mathematics.

Apart from reviewing the original text and correcting a few typos, this second edition incorporates 3 extra chapters, and all 175 illustrations are in colour. I have also extended Chap. 9 on arc length to include parameterisation of curves.

The objective of the book remains the same: to inform the reader about functions and their derivatives, and the inverse process: integration, which can be used for computing area and volume. The emphasis on geometry gives the book relevance to the computer graphics community, and hopefully will provide the mathematical background for professionals working in computer animation, games and allied disciplines to read and understand other books and technical papers where differential and integral notation is found.

The book divides into 16 chapters, with the obligatory Introduction and Conclusion chapters. Chapter 2 reviews the ideas of functions, their notation and the different types encountered in everyday mathematics. This can be skipped by readers already familiar with the subject.

Chapter 3 introduces the idea of limits and derivatives, and how mathematicians have adopted limits in preference to infinitesimals. Most authors introduce integration as a separate subject, but I have included it in this chapter so that it is seen as an antiderivative, rather than something independent.

Chapter 4 looks at derivatives and antiderivatives for a wide range of functions such as polynomial, trigonometric, exponential and logarithmic. It also shows how function sums, products, quotients and function of a function are differentiated.

Chapter 5 covers higher derivatives and how they are used to detect a local maximum and minimum.

Chapter 6 covers partial derivatives, which although are easy to understand, have a reputation for being difficult. This is possibly due to the symbols used, rather than the underlying mathematics. The total derivative is introduced here as it is required in a later chapter.

Chapter 7 introduces the standard techniques for integrating different types of functions. This can be a large subject, and I have deliberately kept the examples simple in order to keep the reader interested and on top of the subject.

Chapter 8 shows how integration reveals the area under a graph and the concept of the Riemann sum. The idea of representing area or volume as the limiting sum of some fundamental unit, is central to understanding Calculus.

Chapter 9 deals with arc length, and uses a variety of worked examples to compute the length of different curves and their parameterisation.

Chapter 10 shows how single and double integrals are used to compute the surface area for different objects. It is also a convenient point to introduce Jacobians, which hopefully I have managed to explain convincingly.

Chapter 11 shows how single, double and triple integrals are used to compute the volume of familiar objects. It also shows how the choice of a coordinate system influences a solution's complexity.

Chapter 12 covers vector-valued functions, and provides a short introduction to this very large subject.

Chapter 13 shows how to calculate tangent and normal vectors for a variety of curves and surfaces, which will be useful in shading algorithms and physically based animation.

Chapter 14 shows how differential Calculus is used to manage geometric continuity in B-splines and Bézier curves.

Chapter 15 looks at the curvature of curves such as a circle, helix, parabola and parametric plane curves. It also shows how to compute the curvature of 2D quadratic and cubic Bézier curves.

I used Springer's excellent author's *LaTeX* development kit on my Apple iMac, which is so fast that I could recompile the entire book in 3 or 4 s, just to change a single character! This book contains over 170 colour illustrations to provide a strong visual interpretation for derivatives, antiderivatives and the calculation of arc length, curvature, tangent vectors, area and volume. I used Apple's *Grapher* application for most of the graphs and rendered images, and *Pages* for the diagrams.

There is no way I could have written this book without the Internet and several excellent books on Calculus. One only has to Google 'What is a Jacobian?' to receive over a 1000 entries in about 0.25 s! YouTube also contains some highly informative presentations on virtually every aspect of Calculus one could want. So I have spent many hours watching, absorbing and disseminating videos, looking for vital pieces of information that are key to understanding a topic.

The books I have referred to include: *Teach Yourself Calculus*, by Hugh Neil, *Calculus of One Variable*, by Keith Hirst, *Inside Calculus*, by George Exner, *Short Calculus*, by Serge Lang and my all-time favourite: *Mathematics from the Birth of Numbers*, by Jan Gullberg. I acknowledge and thank all these authors for the influence they have had on this book. One other book that has helped me is *Digital*

Typography Using LaTeX by Apostolos Syropoulos, Antonis Tsolomitis and Nick Sofroniou.

Writing any book can be a lonely activity, and finding someone willing to read an early draft, and whose opinion one can trust is extremely valuable. Consequently, I thank Dr. Tony Crilly for his valuable feedback after reading the final manuscript. Tony identified flaws in my reasoning and inconsistent notation, and I have incorporated his suggestions. However, I take full responsibility for any mistakes that may have found there way into this publication.

Finally, I would like to thank Helen Desmond, Editor for Computer Science, Springer UK, for her continuing professional support.

Breinton, Herefordshire   Professor Emeritus John Vince
January 2019   M.Tech, Ph.D., D.Sc., C.Eng.

Contents

Chapter 1
Introduction

1.1 What is Calculus?

Well this is an easy question to answer. Basically, Calculus has two parts: *differential* and *integral*. Differential Calculus is used for computing a function's rate of change relative to one of its arguments. Generally, one begins with a function such as $f(x)$, and as x changes, a corresponding change occurs in $f(x)$. *Differentiating* $f(x)$ with respect to x, produces a second function $f'(x)$, which gives the rate of change of $f(x)$ for any x. For example, and without explaining why, if $f(x) = x^2$, then $f'(x) = 2x$, and when $x = 3$, $f(x)$ is changing $2 \times 3 = 6$ times faster than x. Which is rather neat!

In practice, one also writes $y = x^2$, or even $y = f(x)$, which means that differentiating is expressed in a variety of ways:

$$f'(x), \quad \frac{dy}{dx}, \quad \frac{d}{dx}f(x), \quad \frac{d}{dx}y,$$

thus for $y = f(x) = x^2$, we can write

$$f'(x) = 2x, \quad \frac{dy}{dx} = 2x, \quad \frac{d}{dx}f(x) = 2x, \quad \frac{d}{dx}y = 2x.$$

Integral Calculus reverses the operation, where *integrating* $f'(x)$, produces $f(x)$, or something similar. But surely, Calculus can't be as easy as this, you're asking yourself? Well, there are some problems, which is what this book is about. To begin with, not all functions are easily differentiated, as they may contain hidden infinities and discontinuities. Some functions are expressed as products or quotients, and many functions possess more than one argument. All these, and other conditions, must be addressed. Furthermore, integrating a function produces some useful benefits, such as calculating the area under a graph, the length of curves, and the surface area and volume of objects. But more of this later.

© Springer Nature Switzerland AG 2019
J. Vince, *Calculus for Computer Graphics*,
https://doi.org/10.1007/978-3-030-11376-6_1

Fig. 1.1 Two abutting
curves without matching
slopes

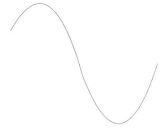

But why should we be interested in rates of change? Well, say we have a function that specifies the changing velocity of an object over time, then differentiating the function gives the rate of change of the function over time, which is the object's acceleration. And knowing the object's mass and acceleration, we can compute the force responsible for the object's acceleration. There are many more reasons for having an interest in rates of change, which will emerge in the following chapters.

1.2   Where is Calculus Used in Computer Graphics?

If you are lucky, you may work in computer graphics without having to use Calculus, but some people have no choice but to understand it, and use it in their work. For example, we often join together curved lines and surfaces. Figure 1.1 shows two abutting curves, where the join is clearly visible. This is because the slope information at the end of the first curve, does not match the slope information at the start of the second curve. By expressing the curves as functions, differentiating them gives their slopes at any point in the form of two other functions. These slope functions can also be differentiated, and by ensuring that the original curves possess the same differentials at the join, a seamless join is created. The same process is used for abutting two or more surface patches.

Calculus finds its way into other aspects of computer graphics such as digital differential analysers (DDAs) for drawing lines and curves, interpolation, curvature, arc-length parametrisation, fluid animation, rendering, animation, modelling, etc. In later chapters I will show how Calculus permits us to calculate surface normals to curves and surfaces, and the curvature of different curves.

1.3   Who Invented Calculus?

More than three-hundred years have passed since the English astronomer, physicist and mathematician Isaac Newton (1643–1727) and the German mathematician Gottfried Leibniz (1646–1716) published their treaties describing Calculus. So called 'infinitesimals' played a pivotal role in early Calculus to determine tangents, area and volume. Incorporating incredibly small quantities (infinitesimals) into a numerical solution, means that products involving them can be ignored, whilst quotients are retained. The final solution takes the form of a ratio representing the change of a function's value, relative to a change in its independent variable.

Although infinitesimal quantities have helped mathematicians for more than two-thousand years solve all sorts of problems, they were not widely accepted as a rigorous mathematical tool. In the latter part of the 19th century, they were replaced by incremental changes that tend towards zero to form a limit identifying some desired result. This was mainly due to the work of the German mathematician Karl Weierstrass (1815–1897), and the French mathematician Augustin-Louis Cauchy (1789–1857).

In spite of the basic ideas of Calculus being relatively easy to understand, it has a reputation for being difficult and intimidating. I believe that the problem lies in the breadth and depth of Calculus, in that it can be applied across a wide range of disciplines, from electronics to cosmology, where the notation often becomes extremely abstract with multiple integrals, multi-dimensional vector spaces and matrices formed from partial differential operators. In this book I introduce the reader to those elements of Calculus that are both easy to understand and relevant to solving various mathematical problems found in computer graphics.

Perhaps you have studied Calculus at some time, and have not had the opportunity to use it regularly and become familiar with its ways, tricks and analytical techniques. In which case, this book could awaken some distant memory and reveal a subject with which you were once familiar. On the other hand, this might be your first journey into the world of functions, limits, differentials and integrals – in which case, you should find the journey exciting!

Chapter 2
Functions

2.1 Introduction

In this chapter the notion of a function is introduced as a tool for generating one numerical quantity from another. In particular, we look at equations, their variables and any possible sensitive conditions. This leads toward the idea of how fast a function changes relative to its independent variable. The second part of the chapter introduces two major operations of Calculus: differentiating, and its inverse, integrating. This is performed without any rigorous mathematical underpinning, and permits the reader to develop an understanding of Calculus without using limits.

2.2 Expressions, Variables, Constants and Equations

One of the first things we learn in mathematics is the construction of *expressions*, such as $2(x + 5) - 2$, using *variables*, *constants* and arithmetic *operators*. The next step is to develop an equation, which is a mathematical statement, in symbols, declaring that two things are exactly the same (or equivalent). For example, (2.1) is the equation representing the surface area of a sphere:

$$S = 4\pi r^2 \tag{2.1}$$

where S and r are variables. They are variables because they take on different values, depending on the size of the sphere. S depends upon the changing value of r, and to distinguish between the two, S is called the *dependent variable*, and r the *independent variable*. Similarly, (2.2) is the equation for the volume of a torus:

$$V = 2\pi^2 r^2 R \tag{2.2}$$

© Springer Nature Switzerland AG 2019
J. Vince, *Calculus for Computer Graphics*,
https://doi.org/10.1007/978-3-030-11376-6_2

where the *dependent variable* V depends on the torus's minor radius r and major radius R, which are both *independent variables*. Note that both formulae include constants 4, π and 2. There are no restrictions on the number of variables or constants employed within an equation.

2.3   Functions

The concept of a function is that of a *dependent relationship*. Some equations merely express an equality, such as $19 = 15 + 4$, but a function is a special type of equation in which the value of one variable (the dependent variable) depends on, and is determined by, the values of one or more other variables (the independent variables). Thus, in the equation

$$S = 4\pi r^2$$

one might say that S is a function of r, and in the equation

$$V = 2\pi^2 r^2 R$$

V is a function of r and also of R.

It is usual to write the independent variables, separated by commas, in brackets immediately after the symbol for the dependent variable, and so the two equations above are usually written

$$S(r) = 4\pi r^2$$

and

$$V(r, R) = 2\pi^2 r^2 R.$$

The order of the independent variables is immaterial.

Mathematically, there is no difference between equations and functions, it is simply a question of notation. However, when we do not have an equation, we can use the idea of a function to help us develop one. For example, no one has been able to find an equation that generates the nth prime number, but I can declare a function $P(n)$ that pretends to perform this operation, such that $P(1) = 2$, $P(2) = 3$, $P(3) = 5$, etc. At least this imaginary function $P(n)$, permits me to move forward and reflect upon its possible inner structure.

A mathematical function *must* have a precise definition. It *must* be predictable, and ideally, work under all conditions.

We are all familiar with mathematical functions such as $\sin x$, $\cos x$, $\tan x$, $\sqrt{x}$, etc., where x is the independent variable. Such functions permit us to confidently write statements such as

$$\sin 30° = 0.5$$
$$\cos 90° = 0.0$$
$$\tan 45° = 1.0$$
$$\sqrt{16} = \pm 4$$

without worrying whether they will always provide a correct answer.

We often need to design a function to perform a specific task. For instance, if I require a function $y(x)$ to compute $x^2 + x + 6$, the independent variable is x and the function is written

$$y(x) = x^2 + x + 6$$

such that

$$y(0) = 0^2 + 0 + 6 = 6$$
$$y(1) = 1^2 + 1 + 6 = 8$$
$$y(2) = 2^2 + 2 + 6 = 12$$
$$y(3) = 3^2 + 3 + 6 = 18.$$

2.3.1 Continuous and Discontinuous Functions

Understandably, a function's value is sensitive to its independent variables. A *simple* square-root function, for instance, expects a positive real number as its independent variable, and registers an error condition for a negative value. On the other hand, a useful square-root function would accept positive and negative numbers, and output a real result for a positive input and a complex result for a negative input.

Another danger condition is the possibility of dividing by zero, which is not permissible in mathematics. For example, the following function $y(x)$ is undefined for $x = 1$, and cannot be displayed on the graph shown in Fig. 2.1.

$$y(x) = \frac{x^2 + 1}{x - 1}$$
$$y(1) = \frac{2}{0}$$

which is why mathematicians include a domain of definition in the specification of a function:

$$y(x) = \frac{x^2 + 1}{x - 1} \quad \text{for } x \neq 1.$$

Fig. 2.1 Graph of
$y = (x^2 + 1)/(x - 1)$
showing the discontinuity at
$x = 1$

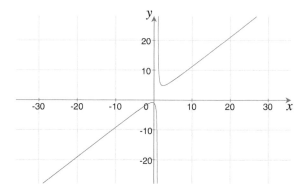

We can create equations or functions that lead to all sorts of mathematical anomalies. For example, (2.3) creates the condition 0/0 when $x = 4$

$$y(x) = \frac{x - 4}{\sqrt{x} - 2} \tag{2.3}$$

$$y(4) = \frac{0}{0}$$

Similarly, mathematicians would write (2.3) as

$$y(x) = \frac{x - 4}{\sqrt{x} - 2} \quad \text{for } x \neq 4.$$

Such conditions have no numerical value. However, this does not mean that these functions are unsound—they are just sensitive to specific values of their independent variable. Fortunately, there is a way of interpreting these results, as we will discover in the next chapter.

2.3.2 Linear Functions

Linear functions are probably the simplest functions we will ever encounter and are based upon equations of the form

$$y = mx + c.$$

Fig. 2.2 Graph of
$y = 0.5x + 2$

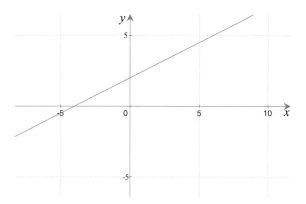

Fig. 2.3 Graph of
$y = 5 \sin x$

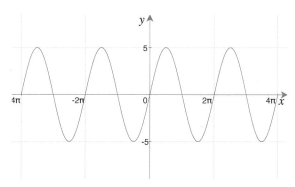

For example, the function $y(x) = 0.5x + 2$ is shown as a graph in Fig. 2.2, where 0.5 is the slope, and 2 is the intercept with the y-axis.

2.3.3 Periodic Functions

Periodic functions are also relatively simple and employ the trigonometric functions sin, cos and tan. For example, the function $y(x) = 5 \sin x$ is shown over the range $-4\pi < x < 4\pi$ as a graph in Fig. 2.3, where the 5 is the amplitude of the sine wave, and x is the angle in radians.

2.3.4 Polynomial Functions

Polynomial functions take the form

$$f(x) = a_n x^n + a_{n-1} x^{n-1} + a_{n-2} x^{n-2} + \cdots b + a_2 x^2 + a_1 x + a_0$$

Fig. 2.4 Graph of $f(x) = 4x^4 - 5x^3 - 8x^2 + 6x - 12$

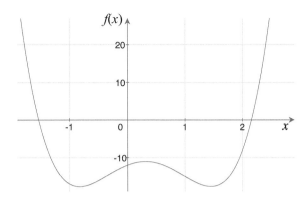

where n takes on some value, and a_n are assorted constants. For example, the function $f(x) = 4x^4 - 5x^3 - 8x^2 + 6x - 12$ is shown in Fig. 2.4.

2.3.5 Function of a Function

In mathematics we often combine functions to describe some relationship succinctly. For example, the trigonometric function

$$f(x) = \sin(2x + 1)$$

is a *function of a function*. Here we have $2x + 1$, which can be expressed as the function

$$u(x) = 2x + 1$$

and the original function becomes

$$f(u(x)) = \sin(u(x)).$$

We can increase the depth of functions to any limit, and in the next chapter we consider how such descriptions are untangled and analysed in Calculus.

2.3.6 Other Functions

You are probably familiar with other functions such as exponential, logarithmic, complex, vector, recursive, etc., which can be combined together to encode simple equations such as

$$e = mc^2$$

or something more difficult such as

$$A(k) = \frac{1}{N} \sum_{j=0}^{N-1} f_j \omega^{-jk} \quad \text{for} \quad k = 0, 1, \ldots, N-1.$$

2.4   A Function's Rate of Change

Mathematicians are particularly interested in the rate at which a function changes relative to its independent variable. Even you would be interested in this characteristic in the context of your salary or pension annuity. For example, I would like to know if my pension fund is growing linearly with time; whether there is some sustained increasing growth rate; or more importantly, if the fund is decreasing! This is what Calculus is about—it enables us to calculate how a function's value changes, relative to its independent variable.

The reason why Calculus appears daunting, is that there is such a wide range of functions to consider: linear, periodic, complex, polynomial, imaginary, rational, exponential, logarithmic, vector, etc. However, we must not be intimidated by such a wide spectrum, as the majority of functions employed in computer graphics are relatively simple, and there are plenty of texts that show how specific functions are tackled.

2.4.1   Slope of a Function

In the linear equation

$$y = mx + c$$

the independent variable is x, but y is also influenced by the constant c, which determines the intercept with the y-axis, and m, which determines the graph's slope. Figure 2.5 shows this equation with 4 different values of m. For any value of x, the slope always equals m, which is what linear means.

In the quadratic equation

$$y = ax^2 + bx + c$$

y is dependent on x, but in a much more subtle way. It is a combination of two components: a square law component ax^2, and a linear component $bx + c$. Figure 2.6 shows these two components and their sum for the equation $y = 0.5x^2 - 2x + 1$.

For any value of x, the slope is different. Figure 2.7 identifies three slopes on the graph. For example, when $x = 2$, $y = -1$, and the slope is zero. When $x = 4$, $y = 1$, and the slope looks as though it equals 2. And when $x = 0$, $y = 1$, the slope looks as though it equals -2.

Fig. 2.5 Graph of
$y = mx + 2$ for different
values of m

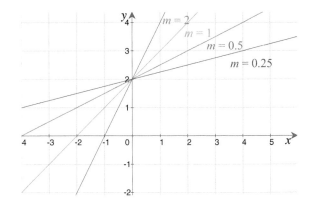

Fig. 2.6 Graph of
$y = 0.5x^2 - 2x + 1$
showing its two components

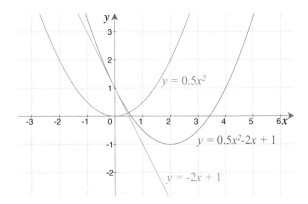

Fig. 2.7 Graph of
$y = 0.5x^2 - 2x + 1$
showing three gradients

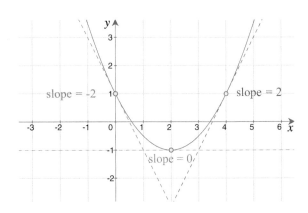

Fig. 2.8 Linear relationship
between slope and x

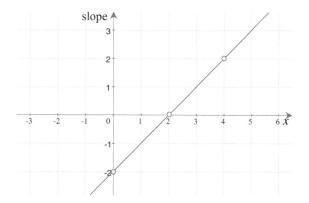

Even though we have only three samples, let's plot the graph of the relationship
between x and the slope m, as shown in Fig. 2.8. Assuming that other values of slope
lie on the same straight line, then the equation relating the slope m to x is

$$m = x - 2.$$

Summarising: we have discovered that the slope of the function

$$f(x) = 0.5x^2 - 2x + 1$$

changes with the independent variable x, and is given by the function

$$f'(x) = x - 2.$$

Note that $f(x)$ is the original function, and $f'(x)$ (pronounced f *prime* of x) is the
function for the slope, which is a convention often used in Calculus.

Remember that we have taken only three sample slopes, and assumed that there
is a linear relationship between the slope and x. Ideally, we should have sampled the
graph at many more points to increase our confidence, but I happen to know that we
are on solid ground!

Calculus enables us to compute the function for the slope from the original function. i.e. to compute $f'(x)$ from $f(x)$:

$$f(x) = 0.5x^2 - 2x + 1 \tag{2.4}$$
$$f'(x) = x - 2. \tag{2.5}$$

Readers who are already familiar with Calculus will know how to compute (2.5)
from (2.4), but for other readers, this is the technique:

1. Take each term of (2.4) in turn and replace ax^n by nax^{n-1}.
2. Therefore $0.5x^2$ becomes x.
3. $-2x$, which can be written $-2x^1$, becomes $-2x^0$, which is -2.
4. 1 is ignored, as it is a constant.
5. Collecting up the terms we have

$$f'(x) = x - 2.$$

This process is called *differentiating* a function, and is easy for this type of polynomial. So easy in fact, we can differentiate the following function without thinking:

$$f(x) = 12x^4 + 6x^3 - 4x^2 + 3x - 8$$
$$f'(x) = 48x^3 + 18x^2 - 8x + 3.$$

This is an amazing relationship, and is one of the reasons why Calculus is so important.

If we can differentiate a polynomial function, surely we can reverse the operation and compute the original function? Well of course! For example, if $f'(x)$ is given by

$$f'(x) = 6x^2 + 4x + 6 \tag{2.6}$$

then this is the technique to compute the original function:

1. Take each term of (2.6) in turn and replace ax^n by $\frac{1}{n+1}ax^{n+1}$.
2. Therefore $6x^2$ becomes $2x^3$.
3. $4x$ becomes $2x^2$.
4. 6 becomes $6x$.
5. Introduce a constant C which might have been present in the original function.
6. Collecting up the terms we have

$$f(x) = 2x^3 + 2x^2 + 6x + C.$$

This process is called *integrating* a function. Thus Calculus is about differentiating and integrating functions, which sounds rather easy, and in some cases it is true. The problem is the breadth of functions that arise in mathematics, physics, geometry, cosmology, science, etc. For example, how do we differentiate or integrate

$$f(x) = \frac{\sin x + \frac{x}{\cosh x}}{\cos^2 x - \ln x^3}?$$

Personally, I don't know, but hopefully, there is a solution somewhere.

Fig. 2.9 A sine curve over the range 0° to 360°

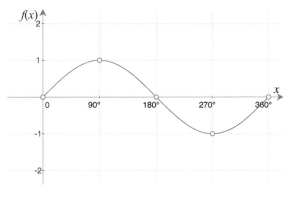

Fig. 2.10 Sampled slopes of a sine curve

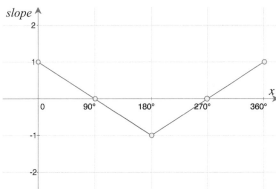

2.4.2 Differentiating Periodic Functions

Now let's try differentiating the sine function by sampling its slope at different points. Figure 2.9 shows a sine curve over the range 0°–360°. When the scales for the vertical and horizontal axes are equal, the slope is 1 at 0° and 360°. The slope is zero at 90° and 270°, and −1 at 180°. Figure 2.10 plots these slope values against x and connects them with straight lines.

It should be clear from Fig. 2.9 that the slope of the sine wave does not change linearly as shown in Fig. 2.10. The slope starts at 1, and for the first 20°, or so, slowly falls away, and then collapses to zero, as shown in Fig. 2.11, which is a cosine wave form. Thus, we can guess that differentiating a sine function creates a cosine function:

$$f(x) = \sin x$$
$$f'(x) = \cos x.$$

Consequently, integrating a cosine function creates a sine function. Now this analysis is far from rigorous, but we will shortly provide one. Before moving on, let's perform a similar 'guesstimate' for the cosine function.

Fig. 2.11 The slope of a
sine curve is a cosine curve

Fig. 2.12 Sampled slopes of
a cosine curve

Fig. 2.13 The slope of a
cosine curve is a negative
sine curve

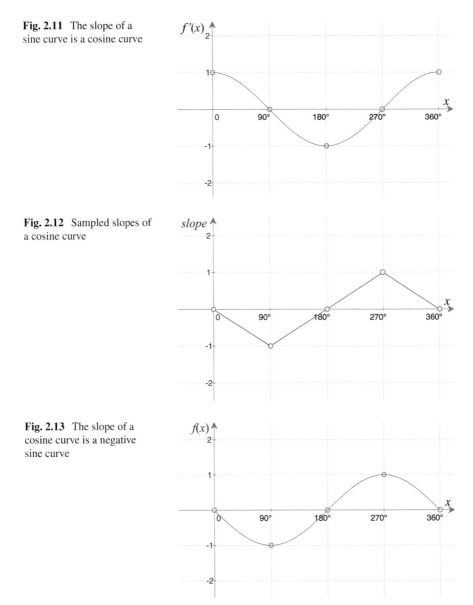

Figure 2.11 shows a cosine curve, where the slope is zero at 0°, 180° and 360°.
The slope equals −1 at 90°, and equals 1 at 270°. Figure 2.12 plots these slope values
against x and connects them with straight lines. Using the same argument for the
sine curve, this can be represented by $f'(x) = -\sin x$ as shown in Fig. 2.13.

Summarising, we have:

$$f(x) = \sin x$$
$$f'(x) = \cos x$$
$$f(x) = \cos x$$
$$f'(x) = -\sin x$$

which illustrates the intimate relationship between the sine and cosine functions.

Just in case you are suspicious of these results, they can be confirmed by differentiating the power series for the sine and cosine functions. For example, the sine and cosine functions are represented by the series

$$\sin x = x - \frac{x^3}{3!} + \frac{x^5}{5!} - \frac{x^7}{7!} + \cdots$$
$$\cos x = 1 - \frac{x^2}{2!} + \frac{x^4}{4!} - \frac{x^6}{6!} + \cdots$$

and differentiating the sine function using the above technique for a polynomial we obtain

$$f'(x) = 1 - \frac{x^2}{2!} + \frac{x^4}{4!} - \frac{x^6}{6!} + \cdots$$

which is the cosine function. Similarly, differentiating the cosine function, we obtain

$$f'(x) = -\left(x - \frac{x^3}{3!} + \frac{x^5}{5!} - \frac{x^7}{7!} + \cdots \right)$$

which is the negative sine function.

Finally, there is a series that when differentiated, remains the same:

$$f(x) = 1 + x + \frac{x^2}{2!} + \frac{x^3}{3!} + \frac{x^4}{4!} + \cdots$$
$$f'(x) = 1 + x + \frac{x^2}{2!} + \frac{x^3}{3!} + \frac{x^4}{4!} + \cdots$$

which is e^x, and has a rate of growth equal to itself!

2.5   Summary

We have covered quite a lot in this chapter, but hopefully it was not too challenging, bearing in mind the subject. We have covered the nature of simple functions and noted that Calculus is interested in a function's rate of change, relative to its independent variable. Differentiating a function creates another function that describes the function's rate of change relative to its independent variable. For simple polynomials, this is a trivial algebraic operation, which can even be undertaken by software. For trigonometric functions, there is a direct relationship between the sine and cosine functions.

Integration is the reverse process, where the original function is derived from a knowledge of the differentiated form. Much more will be said of this process in later chapters.

Chapter 3
Limits and Derivatives

3.1 Introduction

Some quantities, such as the area of a circle or an ellipse, cannot be written precisely, as they incorporate π, which is irrational, but also transcendental; i.e. not a root of a single-variable polynomial whose coefficients are all integers. However, an approximate value can be obtained by devising a definition that includes a parameter that is made infinitesimally small. The techniques of limits and infinitesimals have been used in mathematics for over two-thousand years, and paved the way towards today's Calculus.

Although the principles of integral Calculus were being used by Archimedes (287–212 B.C.) to compute areas, volumes and centres of gravity, it was Isaac Newton and Gottfried Leibniz who are regarded as the true inventors of modern Calculus. Leibniz published his results in 1684, followed by Newton in 1704. However, Newton had been using his *Calculus of fluxions* as early as 1665. Since then, Calculus has evolved conceptually and in notation.

Up until recently, Calculus was described using *infinitesimals*, which are numbers so small, they can be ignored in certain products. However, infinitesimals, no matter how small they are, do not belong to an axiomatic mathematical system, and eventually, Augustin-Louis Cauchy and Karl Weierstrass showed how they could be replaced by limits. In this chapter I show how limits are used to measure a function's rate of change accurately, instead of using intelligent guess work. Limiting conditions also permit us to explore the behaviour of functions that are discontinuous for particular values of their independent variable. For example, rational functions are often sensitive to a specific value of their variable, which gives rise to the meaningless condition $0/0$. Limits permit us to handle such conditions.

We continue to apply limiting conditions to identify a function's derivative, which provides a powerful analytical tool for computing the derivative of function sums, products and quotients. We begin this chapter by exploring small numerical quantities and how they can be ignored if they occur in certain products, but remain important in quotients.

© Springer Nature Switzerland AG 2019
J. Vince, *Calculus for Computer Graphics*,
https://doi.org/10.1007/978-3-030-11376-6_3

3.2 Small Numerical Quantities

The adjective *small* is a relative term, and requires clarification in the context of numbers. For example, if numbers are in the hundreds, and also contain some decimal component, then it seems reasonable to ignore digits after the 3rd decimal place for any quick calculation. For instance,

$$100.000003 \times 200.000006 \approx 20,000$$

and ignoring the decimal part has no significant impact on the general accuracy of the answer, which is measured in tens of thousands.

To develop an algebraic basis for this argument let's divide a number into two parts: a primary part x, and some very small secondary part δx (pronounced *delta x*). In one of the above numbers, $x = 100$ and $\delta x = 0.000003$. Given two such numbers, x_1 and y_1, their product is given by

$$x_1 = x + \delta x$$
$$y_1 = y + \delta y$$
$$x_1 y_1 = (x + \delta x)(y + \delta y)$$
$$= xy + x \cdot \delta y + y \cdot \delta x + \delta x \cdot \delta y.$$

Using $x_1 = 100.000003$ and $y_1 = 200.000006$ we have

$$x_1 y_1 = 100 \times 200 + 100 \times 0.000006 + 200 \times 0.000003 + 0.000003 \times 0.000006$$
$$= 20,000 + 0.0006 + 0.0006 + 0.00000000018$$
$$= 20,000 + 0.0012 + 0.00000000018$$
$$= 20,000.00120000018$$

where it is clear that the products $x \cdot \delta y$, $y \cdot \delta x$ and $\delta x \cdot \delta y$ contribute very little to the result. Furthermore, the smaller we make δx and δy, their contribution becomes even more insignificant. Just imagine if we reduce δx and δy to the level of quantum phenomenon, i.e. 10^{-34}, then their products play no part in every-day numbers. But there is no need to stop there, we can make δx and δy as small as we like, e.g. $10^{-100,000,000,000}$. Later on we employ the device of reducing a number towards zero, such that any products involving them can be dropped from any calculation.

Even though the product of two numbers less than zero is an even smaller number, care must be taken with their quotients. For example, in the above scenario, where $\delta y = 0.000006$ and $\delta x = 0.000003$,

$$\frac{\delta y}{\delta x} = \frac{0.000006}{0.000003} = 2$$

so we must watch out for such quotients.

From now on I will employ the term *derivative* to describe a function's rate of change relative to its independent variable. I will now describe two ways of computing a derivative, and provide a graphical interpretation of the process. The first way uses simple algebraic equations, and the second way uses a functional representation. Needless to say, they both give the same result.

3.3 Equations and Limits

3.3.1 Quadratic Function

Here is a simple algebraic approach using limits to compute the derivative of a quadratic function. Starting with the function $y = x^2$, let x change by δx, and let δy be the corresponding change in y. We then have

$$y = x^2$$
$$y + \delta y = (x + \delta x)^2$$
$$= x^2 + 2x \cdot \delta x + (\delta x)^2$$
$$\delta y = 2x \cdot \delta x + (\delta x)^2.$$

Dividing throughout by δx we have

$$\frac{\delta y}{\delta x} = 2x + \delta x.$$

The ratio $\delta y / \delta x$ provides a measure of how fast y changes relative to x, in increments of δx. For example, when $x = 10$

$$\frac{\delta y}{\delta x} = 20 + \delta x,$$

and if $\delta x = 1$, then $\delta y / \delta x = 21$. Equally, if $\delta x = 0.001$, then $\delta y / \delta x = 20.001$. By making δx smaller and smaller, δy becomes equally smaller, and their ratio converges towards a limiting value of 20.

In this case, as δx approaches zero, $\delta y / \delta x$ approaches $2x$, and is written

$$\lim_{\delta x \to 0} \frac{\delta y}{\delta x} = 2x.$$

Thus in the limit, when $\delta x = 0$, we create a condition where δy is divided by zero— which is a meaningless operation. However, if we hold onto the idea of a limit, as $\delta x \to 0$, it is obvious that the quotient $\delta y / \delta x$ is converging towards $2x$. The subterfuge employed to avoid dividing by zero is to substitute another quotient dy/dx to stand for the limiting condition:

$$\frac{dy}{dx} = \lim_{\delta x \to 0} \frac{\delta y}{\delta x} = 2x.$$

dy/dx (pronounced *dee y dee x*) is the derivative of $y = x^2$, i.e. $2x$. For instance, when $x = 0$, $dy/dx = 0$, and when $x = 3$, $dy/dx = 6$. The derivative dy/dx, is the instantaneous rate at which y changes relative to x.

If we had represented this equation as a function:

$$f(x) = x^2$$

then dy/dx is another way of expressing $f'(x)$.

Now let's introduce two constants into the original quadratic equation to see what effect, if any, they have on the derivative. We begin with

$$y = ax^2 + b$$

and increment x and y:

$$y + \delta y = a(x + \delta x)^2 + b$$
$$= a\left(x^2 + 2x \cdot \delta x + (\delta x)^2\right) + b$$
$$\delta y = a\left(2x \cdot \delta x + (\delta x)^2\right).$$

Dividing throughout by δx:

$$\frac{\delta y}{\delta x} = a(2x + \delta x)$$

and the derivative is

$$\frac{dy}{dx} = \lim_{\delta x \to 0} \frac{\delta y}{\delta x} = 2ax.$$

Thus we see the added constant b disappears (i.e. because it does not change), whilst the multiplied constant a is transmitted through to the derivative.

3.3.2 Cubic Equation

Now let's repeat the above analysis for $y = x^3$:

$$y = x^3$$
$$y + \delta y = (x + \delta x)^3$$
$$= x^3 + 3x^2 \cdot \delta x + 3x(\delta x)^2 + (\delta x)^3$$
$$\delta y = 3x^2 \cdot \delta x + 3x(\delta x)^2 + (\delta x)^3.$$

Dividing throughout by δx:

$$\frac{\delta y}{\delta x} = 3x^2 + 3x \cdot \delta x + (\delta x)^2.$$

Employing the idea of infinitesimals, one would argue that any term involving δx can be ignored, because its numerical value is too small to make any contribution to the result. Similarly, using the idea of limits, one would argue that as δx is made increasingly smaller, towards zero, any term involving δx rapidly disappears.

Using limits, we have

$$\lim_{\delta x \to 0} \frac{\delta y}{\delta x} = 3x^2$$

or

$$\frac{dy}{dx} = \lim_{\delta x \to 0} \frac{\delta y}{\delta x} = 3x^2.$$

We could also show that if $y = ax^3 + b$ then

$$\frac{dy}{dx} = 3ax^2.$$

This incremental technique can be used to compute the derivative of all sorts of functions.

If we continue computing the derivatives of higher-order polynomials, we discover the following pattern:

$$y = x^2, \quad \frac{dy}{dx} = 2x$$
$$y = x^3, \quad \frac{dy}{dx} = 3x^2$$
$$y = x^4, \quad \frac{dy}{dx} = 4x^3$$
$$y = x^5, \quad \frac{dy}{dx} = 5x^4.$$

Clearly, the rule is

$$y = x^n, \quad \frac{dy}{dx} = nx^{n-1}$$

but we need to prove why this is so. The solution is found in the binomial expansion for $(x + \delta x)^n$, which can be divided into three components:

1. Decreasing terms of x.
2. Increasing terms of δx.
3. The terms of Pascal's triangle.

For example, the individual terms of $(x + \delta x)^4$ are:

Decreasing terms of x:	x^4	x^3	x^2	x^1	x^0
Increasing terms of δx:	$(\delta x)^0$	$(\delta x)^1$	$(\delta x)^2$	$(\delta x)^3$	$(\delta x)^4$
The terms of Pascal's triangle:	1	4	6	4	1

which when combined produce

$$x^4 + 4x^3(\delta x) + 6x^2(\delta x)^2 + 4x(\delta x)^3 + (\delta x)^4.$$

Thus when we begin an incremental analysis:

$$y = x^4$$
$$y + \delta y = (x + \delta x)^4$$
$$= x^4 + 4x^3(\delta x) + 6x^2(\delta x)^2 + 4x(\delta x)^3 + (\delta x)^4$$
$$\delta y = 4x^3(\delta x) + 6x^2(\delta x)^2 + 4x(\delta x)^3 + (\delta x)^4.$$

Dividing throughout by δx:

$$\frac{\delta y}{\delta x} = 4x^3 + 6x^2(\delta x)^1 + 4x(\delta x)^2 + (\delta x)^3.$$

In the limit, as δx slides to zero, only the second term of the original binomial expansion remains:

$$4x^3.$$

The second term of the binomial expansion $(1 + \delta x)^n$ is always of the form

$$nx^{n-1}$$

which is the proof we require.

3.3.3   Functions and Limits

In order to generalise the above findings, let's approach the above analysis using a function of the form $y = f(x)$. We begin by noting some arbitrary value of its independent variable and note the function's value. In general terms, this is x and $f(x)$ respectively. We then increase x by a small amount δx, to give $x + \delta x$, and measure the function's value again: $f(x + \delta x)$. The function's change in value is $f(x + \delta x) - f(x)$, whilst the change in the independent variable is δx. The quotient of these two quantities approximates to the function's rate of change at x:

$$\frac{f(x + \delta x) - f(x)}{\delta x}. \tag{3.1}$$

By making δx smaller and smaller towards zero, (3.1) converges towards a limiting value expressed as

$$\frac{dy}{dx} = \lim_{\delta x \to 0} \frac{f(x + \delta x) - f(x)}{\delta x} \tag{3.2}$$

which can be used to compute all sorts of functions. For example, to compute the derivative of $\sin x$ we proceed as follows:

$$y = \sin x$$
$$y + \delta y = \sin(x + \delta x).$$

Using the identity $\sin(A + B) = \sin A \cdot \cos B + \cos A \cdot \sin B$, we have

$$y + \delta y = \sin x \cdot \cos(\delta x) + \cos x \cdot \sin(\delta x)$$
$$\delta y = \sin x \cdot \cos(\delta x) + \cos x \cdot \sin(\delta x) - \sin x$$
$$= \sin x (\cos(\delta x) - 1) + \cos x \cdot \sin(\delta x).$$

Dividing throughout by δx we have

$$\frac{\delta y}{\delta x} = \frac{\sin x}{\delta x} (\cos(\delta x) - 1) + \frac{\sin(\delta x)}{\delta x} \cos x.$$

In the limit as $\delta x \to 0$, $(\cos(\delta x) - 1) \to 0$ and $\sin(\delta x)/\delta x = 1$ (See Appendix A), and

$$\frac{dy}{dx} = \cos x$$

which confirms our 'guesstimate' in Chap. 2. Before moving on, let's compute the derivative of $\cos x$.

$$y = \cos x$$
$$y + \delta y = \cos(x + \delta x).$$

Using the identity $\cos(A + B) = \cos A \cdot \cos B - \sin A \cdot \sin B$, we have

$$y + \delta y = \cos x \cdot \cos(\delta x) - \sin x \cdot \sin(\delta x)$$
$$\delta y = \cos x \cdot \cos(\delta x) - \sin x \cdot \sin(\delta x) - \cos x$$
$$= \cos x (\cos(\delta x) - 1) - \sin x \cdot \sin(\delta x).$$

Dividing throughout by δx we have

$$\frac{\delta y}{\delta x} = \frac{\cos x}{\delta x} (\cos(\delta x) - 1) - \frac{\sin(\delta x)}{\delta x} \sin x.$$

In the limit as $\delta x \to 0$, $(\cos(\delta x) - 1) \to 0$ and $\sin(\delta x)/\delta x = 1$ (See Appendix A), and

$$\frac{dy}{dx} = -\sin x$$

which also confirms our 'guesstimate'. We will continue to employ this strategy to compute the derivatives of other functions later on.

3.3.4   Graphical Interpretation of the Derivative

To illustrate this limiting process graphically, consider the scenario in Fig. 3.1 where the sample point is P. In this case the function is $f(x) = x^2$ and P's coordinates are (x, x^2). We identify another point R, displaced δx to the right of P, with coordinates $(x + \delta x, x^2)$. The point Q on the curve, vertically above R, has coordinates $\left(x + \delta x, (x + \delta x)^2\right)$. When δx is relatively small, the slope of the line PQ approximates to the function's rate of change at P, which is the graph's slope. This is given by

$$
\begin{aligned}
\text{slope} &= \frac{QR}{PR} = \frac{(x + \delta x)^2 - x^2}{\delta x} \\
&= \frac{x^2 + 2x(\delta x) + (\delta x)^2 - x^2}{\delta x} \\
&= \frac{2x(\delta x) + (\delta x)^2}{\delta x} \\
&= 2x + \delta x.
\end{aligned}
$$

We can now reason that as δx is made smaller and smaller, Q approaches P, and slope becomes the graph's slope at P. This is the *limiting* condition:

$$\frac{dy}{dx} = \lim_{\delta x \to 0} (2x + \delta x) = 2x.$$

Fig. 3.1 Sketch of
$f(x) = x^2$

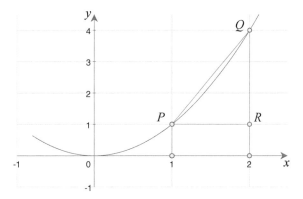

Thus, for any point with coordinates (x, x^2), the slope is given by $2x$. For example, when $x = 0$, the slope is 0, and when $x = 4$, the slope is 8, etc.

3.3.5 Derivatives and Differentials

Given a function $f(x)$, df/dx represents the instantaneous change of f for some x, and is called the *first derivative* of $f(x)$. For linear functions, this is constant, for other functions, the derivative's value changes with x and is represented by a function.

The elements df, dy and dx are called *differentials*, and historically, the derivative used to be called the *differential coefficient*, but has now been dropped in favour of *derivative*. One can see how the idea of a differential coefficient arose if we write, for example:

$$\frac{dy}{dx} = 3x$$

as

$$dy = 3x \, dx.$$

In this case, $3x$ acts like a coefficient of dx, nevertheless, we will use the word *derivative*. It is worth noting that if $y = x$, then $dy/dx = 1$, or $dy = dx$. The two differentials are individual algebraic quantities, which permits us to write statements such as

$$\frac{dy}{dx} = 3x, \qquad dy = 3x \, dx, \qquad dx = \frac{1}{3x} dy.$$

For example, given

$$y = 6x^3 - 4x^2 + 8x + 6$$

then

$$\frac{dy}{dx} = 18x^2 - 8x + 8$$

which is the instantaneous change of y relative to x. When $x = 1$, $dy/dx = 18 - 8 + 8 = 18$, which means that y is changing 18 times faster than x. Consequently, $dx/dy = 1/18$.

Gottfried Leibniz developed what has become known as *Leibniz notation* for differentiation, where

$$\frac{dy}{dx}$$

is a composite definition for a derivative. Leibniz also treated them individually as infinitesimals, which is no longer the case. It was Joseph Lagrange who developed the *prime mark* notation $f'(x)$ to denote the first derivative, with extra prime marks for higher derivatives.

Personally, I find that separating dy/dx into dy and dx has useful pedagogic uses, even if it is not mathematically rigorous!

3.3.6 Integration and Antiderivatives

If it is possible to differentiate a function, it seems reasonable to assume the existence of an inverse process to convert a derivative back to its associated function. Fortunately, this is the case, but there are some limitations. This inverse process is called *integration* and reveals the *antiderivative* of a function. Many functions can be paired together in the form of a derivative and an antiderivative, such as $2x$ with x^2, and $\cos x$ with $\sin x$. However, there are many functions where it is impossible to derive its antiderivative in a precise form. For example, there is no simple, finite functional antiderivative for $\sin(x^2)$ or $(\sin x)/x$. To understand integration, let's begin with a simple derivative.

If we are given

$$\frac{dy}{dx} = 18x^2 - 8x + 8$$

it is not too difficult to reason that the original function could have been

$$y = 6x^3 - 4x^2 + 8x.$$

However, it could have also been

$$y = 6x^3 - 4x^2 + 8x + 2$$

or

$$y = 6x^3 - 4x^2 + 8x + 20$$

or with any other constant. Consequently, when integrating the original function, the integration process has to include a constant:

$$y = 6x^3 - 4x^2 + 8x + C.$$

The value of C is not always required, but it can be determined if we are given some extra information, such as $y = 10$ when $x = 0$, then $C = 10$.

The notation for integration employs a curly 'S' symbol $\int$, which may seem strange, but is short for *sum* and will be explained later. So, starting with

$$\frac{dy}{dx} = 18x^2 - 8x + 8$$

we rewrite this as

$$dy = (18x^2 - 8x + 8)dx$$

and integrate both sides, where dy becomes y and the right-hand-side becomes

$$\int (18x^2 - 8x + 8)\, dx$$

although brackets are not always used:

$$y = \int 18x^2 - 8x + 8\, dx.$$

This equation reads: 'y *is the integral of* $18x^2 - 8x + 8$ *dee x.*' The dx reminds us that x is the independent variable. In this case we can write the answer:

$$\frac{dy}{dx} = 18x^2 - 8x + 8$$
$$dy = 18x^2 - 8x + 8\, dx$$
$$y = \int 18x^2 - 8x + 8\, dx$$
$$= 6x^3 - 4x^2 + 8x + C$$

where C is some constant.

Another example:

$$\frac{dy}{dx} = 6x^2 + 10x$$
$$dy = 6x^2 + 10x \; dx$$
$$y = \int 6x^2 + 10x \; dx$$
$$= 2x^3 + 5x^2 + C.$$

Finally,

$$\frac{dy}{dx} = 1$$
$$dy = 1 \; dx$$
$$y = \int 1 \; dx$$
$$= x + C.$$

The antiderivatives for the sine and cosine functions are written:

$$\int \sin x \; dx = -\cos x + C$$
$$\int \cos x \; dx = \sin x + C$$

which you may think obvious, as we have just computed their derivatives. However, the reason for introducing integration alongside differentiation, is to make you familiar with the notation, and memorise the two distinct processes, as well as lay the foundations for later chapters.

3.4   Summary

This chapter has shown how limits provide a useful tool for computing a function's derivative. Basically, the function's independent variable is disturbed by a very small quantity, typically δx, which alters the function's value. The quotient

$$\frac{f(x + \delta x) - f(x)}{\delta x}$$

is a measure of the function's rate of change relative to its independent variable. By making δx smaller and smaller towards zero, we converge towards a limiting value called the function's derivative. Unfortunately, not all functions possess a derivative, therefore we can only work with functions that can be differentiated. In the next chapter we discover how to differentiate different types of functions and function combinations.

We have also come across integration—the inverse of differentiation—and as we compute the derivatives of other functions, the associated antiderivative will also be included.

3.5 Worked Examples

3.5.1 Limiting Value of a Quotient

Find the limiting value of $\frac{x^8+x^2}{3x^2-x^3}$, as $x \to 0$.

First, we simplify the quotient by dividing the numerator and denominator by x^2:

$$\frac{x^6+1}{3-x}.$$

We can now reason that as $x \to 0$, $(x^6 + 1) \to 1$ and $(3 - x) \to 3$, therefore,

$$\lim_{x \to 0} \frac{x^8 + x^2}{3x^2 - x^3} = \frac{1}{3}$$

which is confirmed by the function's graph in Fig. 3.2.

3.5.2 Limiting Value of a Quotient

Find the limiting value of $\frac{x^2-1}{3x^2-2x-1}$, as $x \to 0$.

Fig. 3.2 Graph of $f(x) = \frac{x^8+x^2}{3x^2-x^3}$

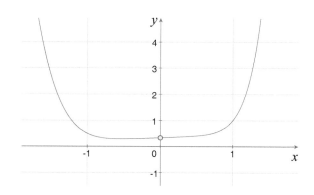

Fig. 3.3 Graph of
$f(x) = \frac{x^2-1}{3x^2-2x-1}$

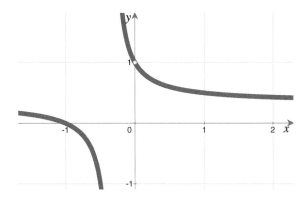

First, we simplify the numerator and denominator:

$$\lim_{x\to 0} \frac{(x+1)(x-1)}{(3x+1)(x-1)} = \lim_{x\to 0} \frac{x+1}{3x+1}.$$

We can now reason that as $x \to 0$, $(x+1) \to 1$ and $(3x+1) \to 1$, therefore,

$$\lim_{x\to 0} \frac{x^2-1}{3x^2-2x-1} = 1$$

which is confirmed by the function's graph in Fig. 3.3.

3.5.3   Derivative

Differentiate $y = 3x^{100} - 4$.
   Using $\frac{dy}{dx} = nx^{n-1}$:

$$\frac{dy}{dx} = 300x^{99}.$$

3.5.4   Slope of a Polynomial

Find the slope of the graph $y = 3x^2 + 2x$ when $x = 2$.

$$\frac{dy}{dx} = 6x + 2.$$

When $x = 2$,

$$\frac{dy}{dx} = 12 + 2 = 14$$

which is the slope.

3.5.5 Slope of a Periodic Function

Find the slope of $y = 6 \sin x$ when $x = \pi/3$.

$$\frac{dy}{dx} = 6 \cos x.$$

When $x = \pi/3$

$$\frac{dy}{dx} = 6 \cos \left(\frac{\pi}{3}\right)$$
$$= 6 \times 0.5 = 3.$$

3.5.6 Integrate a Polynomial

Integrate $dy/dx = 5x^2 + 4x$.

$$dy = 5x^2 + 4x \ dx$$
$$y = \int 5x^2 + 4x \ dx$$
$$= \tfrac{5}{3}x^3 + 2x^2 + C.$$

Chapter 4
Derivatives and Antiderivatives

4.1 Introduction

Mathematical functions come in all sorts of shapes and sizes. Sometimes they are described explicitly where y equals some function of its independent variable(s), such as

$$y = x \sin x$$

or implicitly where y, and its independent variable(s) are part of an equation, such as

$$x^2 + y^2 = 10.$$

A function may reference other functions, such as

$$y = \sin(\cos^2 x)$$

or

$$y = x^{\sin x}.$$

There is no limit to the way functions can be combined, which makes it impossible to cover every eventuality. Nevertheless, in this chapter we explore some useful combinations that prepare us for any future surprises.

In the first section we examine how to differentiate different types of functions, that include sums, products and quotients, which are employed later on to differentiate specific functions such as trigonometric, logarithmic and hyperbolic. Where relevant, I include the appropriate antiderivative to complement its derivative.

© Springer Nature Switzerland AG 2019
J. Vince, *Calculus for Computer Graphics*,
https://doi.org/10.1007/978-3-030-11376-6_4

4.2 Differentiating Groups of Functions

So far, we have only considered simple individual functions, which unfortunately, do not represent the equations found in mathematics, science, physics or even computer graphics. In general, the functions we have to differentiate include sums of functions, functions of functions, function products and function quotients. Let's explore these four scenarios.

4.2.1 Sums of Functions

A function normally computes a numerical value from its independent variable(s), and if it can be differentiated, its derivative generates another function with the same independent variable. Consequently, if a function contains two functions of x, such as u and v, where

$$y = u(x) + v(x)$$

which can be abbreviated to

$$y = u + v$$

then

$$\frac{dy}{dx} = \frac{du}{dx} + \frac{dv}{dx}$$

where we just sum their individual derivatives. For example, let

$$u = 2x^6$$
$$v = 3x^5$$
$$y = u + v$$
$$y = 2x^6 + 3x^5$$

then

$$\frac{dy}{dx} = 12x^5 + 15x^4.$$

Similarly, let

$$u = 2x^6$$
$$v = \sin x$$
$$w = \cos x$$
$$y = u + v + w$$
$$y = 2x^6 + \sin x + \cos x$$

Fig. 4.1 Graph of
$y = 2x^6 + \sin x + \cos x$ and
its derivative,
$y = 12x^5 + \cos x - \sin x$
(dashed)

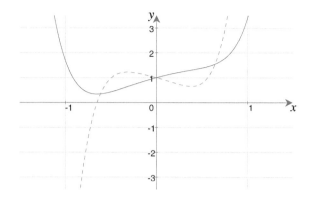

then

$$\frac{dy}{dx} = 12x^5 + \cos x - \sin x.$$

Figure 4.1 shows a graph of $y = 2x^6 + \sin x + \cos x$ and its derivative $y = 12x^5 + \cos x - \sin x$. Differentiating such functions is relatively easy, so too, is integrating. Given

$$\frac{dy}{dx} = \frac{du}{dx} + \frac{dv}{dx}$$

then

$$y = \int u \, dx + \int v \, dx$$
$$= \int (u + v) \, dx$$

and given

$$\frac{dy}{dx} = 12x^5 + \cos x - \sin x.$$

then

$$dy = 12x^5 + \cos x - \sin x \, dx$$
$$y = \int 12x^5 \, dx + \int \cos x \, dx - \int \sin x \, dx$$
$$= 2x^6 + \sin x + \cos x + C.$$

4.2.2   *Function of a Function*

One of the advantages of modern mathematical notation is that it lends itself to unlimited elaboration without introducing any new symbols. For example, the polynomial $3x^2 + 2x$ is easily raised to some power by adding brackets and an appropriate index: $(3x^2 + 2x)^2$. Such an object is a *function of a function*, because the function $3x^2 + 2x$ is subjected to a further squaring function. The question now is: how are such functions differentiated? Well, the answer is relatively easy, but does introduce some new ideas.

Imagine that person A swims twice as fast as person B, who in turn, swims three times as fast as person C. It should be obvious that person A swims six (2×3) times faster than person C. This product rule, also applies to derivatives, because if y changes twice as fast as u, i.e. $\frac{dy}{du} = 2$, and u changes three times as fast as x, i.e. $\frac{du}{dx} = 3$, then y changes six times as fast as x:

$$\frac{dy}{dx} = \frac{dy}{du} \cdot \frac{du}{dx}.$$

To differentiate

$$y = (3x^2 + 2x)^2$$

we substitute

$$u = 3x^2 + 2x$$

then

$$y = u^2$$

and

$$\frac{dy}{du} = 2u$$
$$= 2(3x^2 + 2x)$$
$$= 6x^2 + 4x.$$

Next, we require $\frac{du}{dx}$:

$$u = 3x^2 + 2x$$
$$\frac{du}{dx} = 6x + 2$$

therefore, we can write

Fig. 4.2 Graph of
$y = (3x^2 + 2x)^2$ and its
derivative,
$\frac{dy}{dx} = 36x^3 + 36x^2 + 8x$
(dashed)

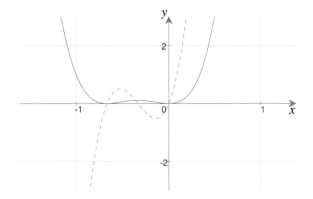

$$\frac{dy}{dx} = \frac{dy}{du} \cdot \frac{du}{dx}$$
$$= (6x^2 + 4x)(6x + 2)$$
$$= 36x^3 + 36x^2 + 8x.$$

This result is easily verified by expanding the original polynomial and differentiating:

$$y = (3x^2 + 2x)^2$$
$$= (3x^2 + 2x)(3x^2 + 2x)$$
$$= 9x^4 + 12x^3 + 4x^2$$
$$\frac{dy}{dx} = 36x^3 + 36x^2 + 8x.$$

Figure 4.2 shows a graph of $y = (3x^2 + 2x)^2$ and its derivative, $\frac{dy}{dx} = 36x^3 + 36x^2 + 8x$.

To differentiate $y = \sin(ax)$, which is a function of a function, we proceed as follows:

$$y = \sin(ax).$$

Substitute u for ax:

$$y = \sin u$$
$$\frac{dy}{du} = \cos u$$
$$= \cos(ax).$$

Next, we require $\frac{du}{dx}$:

$$u = ax$$
$$\frac{du}{dx} = a$$

therefore, we can write

$$\frac{dy}{dx} = \frac{dy}{du} \cdot \frac{du}{dx}$$
$$= \cos(ax)a$$
$$= a\cos(ax).$$

Consequently, given

$$\frac{dy}{dx} = \cos(ax)$$

then

$$y = \int \cos(ax)\, dx$$
$$= \frac{1}{a}\sin(ax) + C.$$

Similarly, given

$$\frac{dy}{dx} = \sin(ax)$$

then

$$y = \int \sin(ax)\, dx$$
$$= -\frac{1}{a}\cos(ax) + C.$$

The equation $y = \sin(x^2)$ is also a function of a function, and is differentiated as follows:

$$y = \sin(x^2).$$

Substitute u for x^2:

$$y = \sin u$$
$$\frac{dy}{du} = \cos u$$
$$= \cos(x^2).$$

Next, we require $\frac{du}{dx}$:

Fig. 4.3 Graph of $y = \sin(x^2)$ and its derivative, $\frac{dy}{dx} = 2x\cos(x^2)$ (dashed)

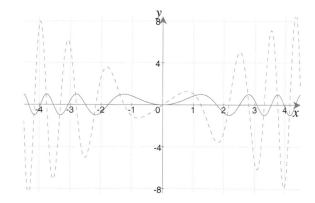

$$u = x^2$$

$$\frac{du}{dx} = 2x$$

therefore, we can write

$$\frac{dy}{dx} = \frac{dy}{du} \cdot \frac{du}{dx}$$
$$= \cos(x^2)2x$$
$$= 2x\cos(x^2).$$

Figure 4.3 shows a graph of $y = \sin(x^2)$ and its derivative, $\frac{dy}{dx} = 2x\cos(x^2)$. In general, there can be any depth of functions within a function, which permits us to write the *chain rule* for derivatives:

$$\frac{dy}{dx} = \frac{dy}{du} \cdot \frac{du}{dv} \cdot \frac{dv}{dw} \cdot \frac{dw}{dx}$$

4.2.3 Function Products

Function products occur frequently in every-day mathematics, and involve the product of two, or more functions. Here are three simple examples:

$$y = (3x^2 + 2x)(2x^2 + 3x)$$
$$y = \sin x \cdot \cos x$$
$$y = x^2 \sin x.$$

When it comes to differentiating function products of the form

$$y = uv$$

it seems natural to assume that

$$\frac{dy}{dx} = \frac{du}{dx} \cdot \frac{dv}{dx} \tag{4.1}$$

which, unfortunately, is incorrect. For example, in the case of

$$y = (3x^2 + 2x)(2x^2 + 3x)$$

differentiating using (4.1) produces

$$\frac{dy}{dx} = (6x + 2)(4x + 3)$$
$$= 24x^2 + 26x + 6.$$

However, if we expand the original product and then differentiate, we obtain

$$y = (3x^2 + 2x)(2x^2 + 3x)$$
$$= 6x^4 + 13x^3 + 6x^2$$
$$\frac{dy}{dx} = 24x^3 + 39x^2 + 12x$$

which is correct, but differs from the first result. Obviously, (4.1) must be wrong. So let's return to first principles and discover the correct rule.

So far, we have incremented the independent variable—normally x—by δx to discover the change in y—normally δy. Next, we see how the same notation can be used to increment functions.

Given the following functions of x, u and v, where

$$y = uv$$

if x increases by δx, then there will be corresponding changes of δu, δv and δy, in u, v and y respectively. Therefore,

$$y + \delta y = (u + \delta u)(v + \delta v)$$
$$= uv + u\delta v + v\delta u + \delta u \cdot \delta v$$
$$\delta y = u\delta v + v\delta u + \delta u \cdot \delta v.$$

Dividing throughout by δx we have

$$\frac{\delta y}{\delta x} = u\frac{\delta v}{\delta x} + v\frac{\delta u}{\delta x} + \delta u\frac{\delta v}{\delta x}.$$

In the limiting condition:

$$\frac{dy}{dx} = \lim_{\delta x \to 0} \left(u \frac{\delta v}{\delta x} \right) + \lim_{\delta x \to 0} \left(v \frac{\delta u}{\delta x} \right) + \lim_{\delta x \to 0} \left(\delta u \frac{\delta v}{\delta x} \right).$$

As $\delta x \to 0$, then $\delta u \to 0$ and $\left(\delta u \frac{\delta v}{\delta x} \right) \to 0$. Therefore,

$$\frac{dy}{dx} = u \frac{dv}{dx} + v \frac{du}{dx}. \qquad (4.2)$$

Using (4.2) for the original function product:

$$u = 3x^2 + 2x$$
$$v = 2x^2 + 3x$$
$$y = uv$$
$$\frac{du}{dx} = 6x + 2$$
$$\frac{dv}{dx} = 4x + 3$$
$$\frac{dy}{dx} = u \frac{dv}{dx} + v \frac{du}{dx}$$
$$= (3x^2 + 2x)(4x + 3) + (2x^2 + 3x)(6x + 2)$$
$$= (12x^3 + 17x^2 + 6x) + (12x^3 + 22x^2 + 6x)$$
$$= 24x^3 + 39x^2 + 12x$$

which agrees with our previous prediction. Figure 4.4 shows the graph of $y = (3x^2 + 2x)(2x^2 + 3x)$ and its derivative, $\frac{dy}{dx} = 24x^3 + 39x^2 + 12x$.

The equation $y = \sin x \cos x$ contains the product of two functions and is differentiated using (4.2) as follows:

Fig. 4.4 Graph of $y = (3x^2 + 2x)(2x^2 + 3x)$ and its derivative, $\frac{dy}{dx} = 24x^3 + 39x^2 + 12x$ (dashed)

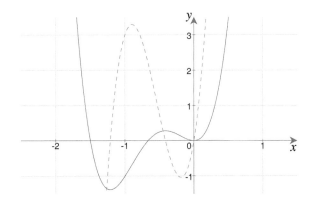

$$y = \sin x \cdot \cos x$$
$$u = \sin x$$
$$\frac{du}{dx} = \cos x$$
$$v = \cos x$$
$$\frac{dv}{dx} = -\sin x$$
$$\frac{dy}{dx} = u\frac{dv}{dx} + v\frac{du}{dx}$$
$$= \sin x(-\sin x) + \cos x \cdot \cos x$$
$$= \cos^2 x - \sin^2 x$$
$$= \cos(2x).$$

Using the identity $\sin(2x) = 2 \sin x \cdot \cos x$, we can rewrite the original function as

$$y = \sin x \cdot \cos x$$
$$\frac{dy}{dx} = \tfrac{1}{2} \sin(2x)$$
$$= \cos(2x)$$

which confirms the above derivative. Now let's consider the antiderivative of $\cos(2x)$. Given

$$\frac{dy}{dx} = \cos(2x)$$

then

$$y = \int \cos(2x)\, dx$$
$$= \tfrac{1}{2} \sin(2x) + C$$
$$= \sin x \cdot \cos x + C.$$

Figure 4.5 shows the graph of $y = \sin x \cdot \cos x$ and its derivative, $\frac{dy}{dx} = \cos(2x)$.

4.2.4   Function Quotients

Next, we investigate how to differentiate the quotient of two functions. We begin with two functions of x, u and v, where

$$y = \frac{u}{v}$$

Fig. 4.5 Graph of
$y = \sin x \cdot \cos x$ and its
derivative, $\frac{dy}{dx} = \cos 2x$
(dashed)

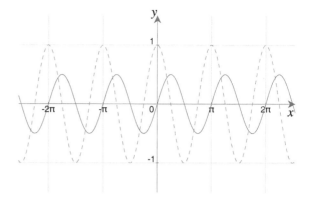

which makes y also a function of x.

We now increment x by δx and measure the change in u as δu, and the change in v as δv. Consequently, the change in y is δy:

$$y + \delta y = \frac{u + \delta u}{v + \delta v}$$

$$\delta y = \frac{u + \delta u}{v + \delta v} - \frac{u}{v}$$

$$= \frac{v(u + \delta u) - u(v + \delta v)}{v(v + \delta v)}$$

$$= \frac{v\delta u - u\delta v}{v(v + \delta v)}.$$

Dividing throughout by δx we have

$$\frac{\delta y}{\delta x} = \frac{v\dfrac{\delta u}{\delta x} - u\dfrac{\delta v}{\delta x}}{v(v + \delta v)}.$$

As $\delta x \to 0$, δu, δv and δy also tend towards zero, and the limiting conditions are

$$\frac{dy}{dx} = \lim_{\delta x \to 0} \frac{\delta y}{\delta x}$$

$$v\frac{du}{dx} = \lim_{\delta x \to 0} v\frac{\delta u}{\delta x}$$

$$u\frac{dv}{dx} = \lim_{\delta x \to 0} u\frac{\delta v}{\delta x}$$

$$v^2 = \lim_{\delta x \to 0} v(v + \delta v)$$

therefore,

Fig. 4.6 Graph of $y =$
$(x^2 + 3)(x + 2)/(x^2 + 3)$
and its derivative, $\frac{dy}{dx} = 1$
(dashed)

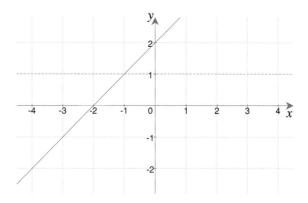

$$\frac{dy}{dx} = \frac{v\dfrac{du}{dx} - u\dfrac{dv}{dx}}{v^2}.$$

As an example, let's differentiate

$$y = \frac{x^3 + 2x^2 + 3x + 6}{x^2 + 3}.$$

Substitute $u = x^3 + 2x^2 + 3x + 6$ and $v = x^2 + 3$, then

$$\frac{du}{dx} = 3x^2 + 4x + 3$$

$$\frac{dv}{dx} = 2x$$

$$\frac{dy}{dx} = \frac{(x^2 + 3)(3x^2 + 4x + 3) - (x^3 + 2x^2 + 3x + 6)(2x)}{(x^2 + 3)^2}$$

$$= \frac{(3x^4 + 4x^3 + 3x^2 + 9x^2 + 12x + 9) - (2x^4 + 4x^3 + 6x^2 + 12x)}{x^4 + 6x^2 + 9}$$

$$= \frac{x^4 + 6x^2 + 9}{x^4 + 6x^2 + 9}$$

$$= 1$$

which is not a surprising result when one sees that the original function has the factors

$$y = \frac{(x^2 + 3)(x + 2)}{x^2 + 3} = x + 2$$

whose derivative is 1. Figure 4.6 shows a graph of $y = (x^2 + 3)(x + 2)/(x^2 + 3)$
and its derivative, $\frac{dy}{dx} = 1$.

Table 4.1 Rules for differentiating function combinations

Function	dy/dx
$y = u(x) \pm v(x)$	$\dfrac{du}{dx} \pm \dfrac{dv}{dx}$
$y = u(v(x))$	$\dfrac{dy}{du} \cdot \dfrac{du}{dx}$
$y = u(x) \cdot v(x)$	$u\dfrac{dv}{dx} + v\dfrac{du}{dx}$
$y = u(x)/v(x)$	$\dfrac{v\dfrac{du}{dx} - u\dfrac{dv}{dx}}{v^2}$

4.2.5 Summary: Groups of Functions

Table 4.1 shows the rules for differentiating function sums, products, quotients and function of a function.

4.3 Differentiating Implicit Functions

Functions conveniently fall into two types: explicit and implicit. An explicit function, describes a function in terms of its independent variable(s), such as

$$y = a \sin x + b \cos x$$

where the value of y is determined by the values of a, b and x. On the other hand, an implicit function, such as

$$x^2 + y^2 = 25$$

combines the function's name with its definition. In this case, it is easy to untangle the explicit form:

$$y = \sqrt{25 - x^2}.$$

So far, we have only considered differentiating explicit functions, so now let's examine how to differentiate implicit functions. Let's begin with a simple explicit function and differentiate it as it is converted into its implicit form.

Let

$$y = 2x^2 + 3x + 4$$

then

$$\frac{dy}{dx} = 4x + 3.$$

Now let's start the conversion into the implicit form by bringing the constant 4 over to the left-hand side:

$$y - 4 = 2x^2 + 3x$$

differentiating both sides:

$$\frac{dy}{dx} = 4x + 3.$$

Bringing 4 and $3x$ across to the left-hand side:

$$y - 3x - 4 = 2x^2$$

differentiating both sides:

$$\frac{dy}{dx} - 3 = 4x$$

$$\frac{dy}{dx} = 4x + 3.$$

Finally, we have

$$y - 2x^2 - 3x - 4 = 0$$

differentiating both sides:

$$\frac{dy}{dx} - 4x - 3 = 0$$

$$\frac{dy}{dx} = 4x + 3$$

which seems straight forward.

The reason for working through this example is to remind us that when y is differentiated we get dy/dx. Consequently, the following examples should be understood:

$$y + \sin x + 4x = 0$$

$$\frac{dy}{dx} + \cos x + 4 = 0$$

$$\frac{dy}{dx} = -\cos x - 4.$$

$$y + x^2 - \cos x = 0$$

$$\frac{dy}{dx} + 2x + \sin x = 0$$

$$\frac{dy}{dx} = -2x - \sin x.$$

But how do we differentiate $y^2 + x^2 = r^2$? Well, the important difference between this implicit function and previous functions, is that it involves a function of a function. y is not only a function of x, but is squared, which means that we must employ the chain rule described earlier:

$$\frac{dy}{dx} = \frac{dy}{du} \cdot \frac{du}{dx}.$$

Therefore, given

$$y^2 + x^2 = r^2$$

$$2y\frac{dy}{dx} + 2x = 0$$

$$\frac{dy}{dx} = \frac{-2x}{2y}$$

$$= \frac{-x}{\sqrt{r^2 - x^2}}.$$

This is readily confirmed by expressing the original function in its explicit form and differentiating:

$$y = (r^2 - x^2)^{\frac{1}{2}}$$

which is a function of a function.

Let $u = r^2 - x^2$, then

$$\frac{du}{dx} = -2x.$$

As $y = u^{\frac{1}{2}}$, then

$$\frac{dy}{du} = \frac{1}{2}u^{-\frac{1}{2}}$$

$$= \frac{1}{2u^{\frac{1}{2}}}$$

$$= \frac{1}{2\sqrt{r^2 - x^2}}.$$

However,

$$\frac{dy}{dx} = \frac{dy}{du} \cdot \frac{du}{dx}$$

$$= \frac{-2x}{2\sqrt{r^2 - x^2}}$$

$$= \frac{-x}{\sqrt{r^2 - x^2}}$$

which agrees with the implicit differentiated form.

As an another example, let's find dy/dx for

$$x^2 - y^2 + 4x = 6y.$$

Differentiating, we have

$$2x - 2y\frac{dy}{dx} + 4 = 6\frac{dy}{dx}.$$

Rearranging the terms, we have

$$2x + 4 = 6\frac{dy}{dx} + 2y\frac{dy}{dx}$$
$$= \frac{dy}{dx}(6 + 2y)$$
$$\frac{dy}{dx} = \frac{2x + 4}{6 + 2y}.$$

If, for example, we have to find the slope of $x^2 - y^2 + 4x = 6y$ at the point $(4, 3)$, then we simply substitute $x = 4$ and $y = 3$ in dy/dx to obtain the answer 1.

Finally, let's differentiate $x^n + y^n = a^n$:

$$x^n + y^n = a^n$$
$$nx^{n-1} + ny^{n-1}\frac{dy}{dx} = 0$$
$$\frac{dy}{dx} = -\frac{nx^{n-1}}{ny^{n-1}}$$
$$\frac{dy}{dx} = -\frac{x^{n-1}}{y^{n-1}}.$$

4.4   Differentiating Exponential and Logarithmic Functions

4.4.1   Exponential Functions

Exponential functions have the form $y = a^x$, where the independent variable is the exponent. Such functions are used to describe various forms of growth or decay, from the compound interest law, to the rate at which a cup of tea cools down. One special value of a is 2.718 282.., called e, where

$$e = \lim_{n \to \infty} \left(1 + \frac{1}{n}\right)^n.$$

Fig. 4.7 Graphs of $y = e^x$
and $y = e^{-x}$

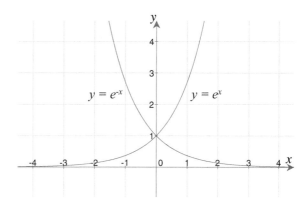

Raising e to the power x:

$$e^x = \lim_{n \to \infty} \left(1 + \frac{1}{n}\right)^{nx}$$

which, using the Binomial Theorem, is

$$e^x = 1 + x + \frac{x^2}{2!} + \frac{x^3}{3!} + \frac{x^4}{4!} + \cdots .$$

If we let

$$y = e^x$$

$$\frac{dy}{dx} = 1 + x + \frac{x^2}{2!} + \frac{x^3}{3!} + \frac{x^4}{4!} + \cdots$$

$$= e^x.$$

which is itself. Figure 4.7 shows graphs of $y = e^x$ and $y = e^{-x}$.

Now let's differentiate $y = a^x$. We know from the rules of logarithms that

$$\log x^n = n \log x$$

therefore, given

$$y = a^x$$

then taking natural logarithms:

$$\ln y = \ln a^x = x \ln a$$

therefore

$$y = e^{x \ln a}$$

which means that

$$a^x = e^{x \ln a}.$$

Consequently,

$$\frac{d}{dx}a^x = \frac{d}{dx}e^{x \ln a}$$
$$= \ln a \cdot e^{x \ln a}$$
$$= \ln a \cdot a^x.$$

Similarly, it can be shown that

$$y = e^{-x}, \quad \frac{dy}{dx} = -e^{-x}$$

$$y = e^{ax}, \quad \frac{dy}{dx} = ae^{ax}$$

$$y = e^{-ax}, \quad \frac{dy}{dx} = -ae^{-ax}$$

$$y = a^x, \quad \frac{dy}{dx} = \ln a \cdot a^x$$

$$y = a^{-x}, \quad \frac{dy}{dx} = -\ln a \cdot a^{-x}.$$

The exponential antiderivatives are written:

$$\int e^x \, dx = e^x + C$$

$$\int e^{-x} \, dx = -e^{-x} + C$$

$$\int e^{ax} \, dx = \frac{1}{a}e^{ax} + C$$

$$\int e^{-ax} \, dx = -\frac{1}{a}e^{ax} + C$$

$$\int a^x \, dx = \frac{1}{\ln a}a^x + C$$

$$\int a^{-x} \, dx = -\frac{1}{\ln a}a^{-x} + C.$$

4.4.2 Logarithmic Functions

Given a function of the form

$$y = \ln x$$

Fig. 4.8 Graph of $y = \ln x$ and its derivative, $\frac{dy}{dx} = \frac{1}{x}$ (dashed)

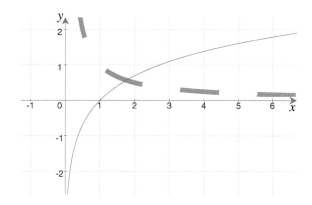

then
$$x = e^y.$$

Therefore,
$$\frac{dx}{dy} = e^y$$
$$= x$$
$$\frac{dy}{dx} = \frac{1}{x}.$$

Thus
$$\frac{d}{dx} \ln x = \frac{1}{x}.$$

Figure 4.8 shows the graph of $y = \ln x$ and its derivative, $\frac{dy}{dx} = \frac{1}{x}$. Conversely,

$$\int \frac{1}{x} \, dx = \ln |x| + C.$$

When differentiating logarithms to a base a, we employ the conversion formula:

$$y = \log_a x$$
$$= \ln x \cdot \log_a e$$

whose derivative is
$$\frac{dy}{dx} = \frac{1}{x} \log_a e.$$

When $a = 10$, then $\log_{10} e = 0.434\,3\ldots$ and

Fig. 4.9 Graph of
$y = \log_{10} x$ and its
derivative, $\frac{dy}{dx} \approx \frac{0.4343}{x}$
(dashed)

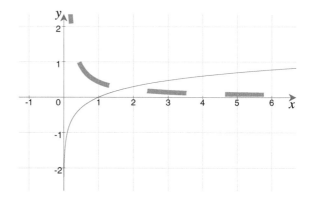

Table 4.2 Rules for differentiating exponential and logarithmic functions

$f(x)$	dy/dx
e^x	e^x
e^{-x}	$-e^{-x}$
e^{ax}	ae^{ax}
e^{-ax}	$-ae^{ax}$
a^x	$\ln a \cdot a^x$
a^{-x}	$-\ln a \cdot a^{-x}$
$\ln x$	$\frac{1}{x}$
$\log_a x$	$\frac{1}{x} \log_a e$
$\log_{10} x$	$\approx \frac{0.4343}{x}$

$$\frac{d}{dx} \log_{10} x \approx \frac{1}{x} 0.4343$$

Figure 4.9 shows the graph of $y = \log_{10} x$ and its derivative, $\frac{dy}{dx} \approx \frac{0.4343}{x}$.

4.4.3   Summary: Exponential and Logarithmic Functions

Table 4.2 shows the rules for differentiating exponential and logarithmic functions, and Table 4.3 shows the rules for integrating exponential functions.

4.5   Differentiating Trigonometric Functions

So far, we have only differentiated two trigonometric functions: $\sin x$ and $\cos x$, so let's add $\tan x$, $\csc x$, $\sec x$ and $\cot x$ to the list, as well as their inverse forms.

Table 4.3 Rules for integrating exponential functions

$f(x)$	$\int f(x)\,dx$
e^x	$e^x + C$
e^{-x}	$-e^{-x} + C$
e^{ax}	$\frac{1}{a}e^{ax} + C$
e^{-ax}	$-\frac{1}{a}e^{-ax} + C$
a^x	$\frac{1}{\ln a}a^x + C$
a^{-x}	$-\frac{1}{\ln a}a^{-x} + C$

4.5.1 Differentiating tan

Rather than return to first principles and start incrementing x by δx, we can employ the rules for differentiating different function combinations and various trigonometric identities. In the case of $\tan(ax)$, this can be written as

$$\tan(ax) = \frac{\sin(ax)}{\cos(ax)}$$

and employ the quotient rule:

$$\frac{dy}{dx} = \frac{v\dfrac{du}{dx} - u\dfrac{dv}{dx}}{v^2}.$$

Therefore, let $u = \sin(ax)$ and $v = \cos(ax)$, and

$$\frac{dy}{dx} = \frac{a\cos(ax) \cdot \cos(ax) + a\sin(ax) \cdot \sin(ax)}{\cos^2(ax)}$$

$$= \frac{a(\cos^2(ax) + \sin^2(ax))}{\cos^2(ax)}$$

$$= \frac{a}{\cos^2(ax)}$$

$$= a(1 + \tan^2(ax)).$$

Figure 4.10 shows a graph of $y = \tan x$ and its derivative $y = 1 + \tan^2 x$. It follows that

$$\int \sec^2(ax)\,dx = \frac{1}{a}\tan(ax) + C.$$

Fig. 4.10 Graph of $y = \tan x$ and its derivative $y = 1 + \tan^2 x$ (dashed)

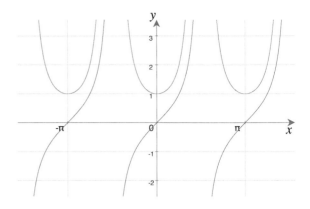

4.5.2 Differentiating csc

Using the quotient rule:

$$y = \csc(ax)$$

$$= \frac{1}{\sin(ax)}$$

$$\frac{dy}{dx} = \frac{0 - a\cos(ax)}{\sin^2(ax)}$$

$$= \frac{-a\cos(ax)}{\sin^2(ax)}$$

$$= -\frac{a}{\sin(ax)} \cdot \frac{\cos(ax)}{\sin(ax)}$$

$$= -a\csc(ax) \cdot \cot(ax).$$

Figure 4.11 shows a graph of $y = \csc x$ and its derivative $y = -\csc x \cdot \cot x$. It follows that

$$\int \csc(ax) \cdot \cot(ax)\, dx = -\frac{1}{a}\csc(ax) + C.$$

4.5.3 Differentiating sec

Using the quotient rule:

$$y = \sec(ax)$$

Fig. 4.11 Graph of
$y = \csc x$ and its derivative
$y = -\csc x \cot x$ (dashed)

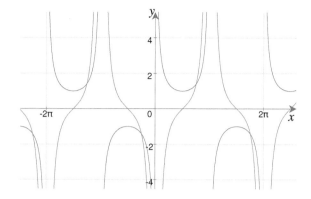

Fig. 4.12 Graph of
$y = \sec x$ and its derivative
$y = \sec x \tan x$ (dashed)

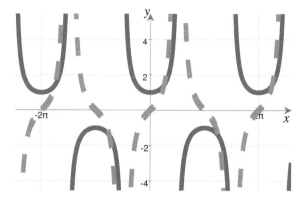

$$
\frac{dy}{dx} =
\begin{aligned}
&= \frac{1}{\cos(ax)} \\
&= \frac{-(-a\sin(ax))}{\cos^2(ax)} \\
&= \frac{a\sin(ax)}{\cos^2(ax)} \\
&= \frac{a}{\cos(ax)} \cdot \frac{\sin(ax)}{\cos(ax)} \\
&= a\sec(ax) \cdot \tan(ax).
\end{aligned}
$$

Figure 4.12 shows a graph of $y = \csc x$ and its derivative $y = -\csc x \cdot \cot x$.
It follows that

$$
\int \sec(ax) \cdot \tan(ax)\,dx = \tfrac{1}{a}\sec(ax) + C.
$$

Fig. 4.13 Graph of
$y = \cot x$ and its derivative,
$\frac{dy}{dx} = -(1 + \cot^2 x)$
(dashed)

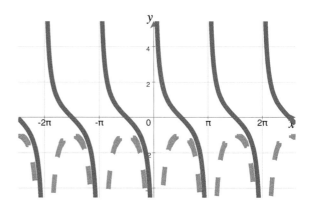

4.5.4   Differentiating cot

Using the quotient rule:

$$y = \cot(ax)$$

$$= \frac{1}{\tan(ax)}$$

$$\frac{dy}{dx} = \frac{-a\sec^2(ax)}{\tan^2(ax)}$$

$$= -\frac{a}{\cos^2(ax)} \cdot \frac{\cos^2(ax)}{\sin^2(ax)}$$

$$= -\frac{a}{\sin^2(ax)}$$

$$= -a\csc^2(ax)$$

$$= -a(1 + \cot^2(ax)).$$

Figure 4.13 shows a graph of $y = \cot x$ and its derivative $y = -(1 + \cot^2 x)$.
    It follows that

$$\int \csc^2(ax)\, dx = -\tfrac{1}{a}\cot(ax) + C.$$

4.5.5   Differentiating arcsin, arccos and arctan

These inverse functions are solved using a clever strategy.

Let

$$x = \sin y$$

then

$$y = \arcsin x.$$

Differentiating the first expression, we have

$$\frac{dx}{dy} = \cos y$$

$$\frac{dy}{dx} = \frac{1}{\cos y}$$

and as $\sin^2 y + \cos^2 y = 1$, then

$$\cos y = \sqrt{1 - \sin^2 y} = \sqrt{1 - x^2}$$

and

$$\frac{d}{dx} \arcsin x = \frac{1}{\sqrt{1 - x^2}}.$$

Using a similar technique, it can be shown that

$$\frac{d}{dx} \arccos x = -\frac{1}{\sqrt{1 - x^2}}$$

$$\frac{d}{dx} \arctan x = \frac{1}{1 + x^2}.$$

It follows that

$$\int \frac{dx}{\sqrt{1 - x^2}} = \arcsin x + C$$

$$\int \frac{dx}{1 + x^2} = \arctan x + C.$$

4.5.6 Differentiating arccsc, arcsec and arccot

Let

$$y = \text{arccsc } x$$

then

$$x = \csc y$$

$$= \frac{1}{\sin y}$$

$$\frac{dx}{dy} = \frac{-\cos y}{\sin^2 y}$$

$$\frac{dy}{dx} = \frac{-\sin^2 y}{\cos y}$$

$$= -\frac{1}{x^2} \frac{x}{\sqrt{x^2 - 1}}$$

$$\frac{d}{dx}\text{arccsc } x = -\frac{1}{x\sqrt{x^2 - 1}}.$$

Similarly,

$$\frac{d}{dx}\text{arcsec } x = \frac{1}{x\sqrt{x^2 - 1}}$$

$$\frac{d}{dx}\text{arccot } x = -\frac{1}{x^2 + 1}.$$

It follows:

$$\int \frac{dx}{x\sqrt{x^2 - 1}} = \text{arcsec } |x| + C$$

$$\int \frac{dx}{x^2 + 1} = -\text{arccot } x + C.$$

4.5.7   Summary: Trigonometric Functions

Table 4.4 shows the rules for differentiating trigonometric functions, and Table 4.5 shows the rules for differentiating inverse trigonometric functions.

Table 4.6 shows the rules for integrating trigonometric functions, and Table 4.7 shows the rules for integrating inverse trigonometric functions.

Table 4.4 The rules for differentiating trigonometric functions

y	dy/dx
$\sin(ax)$	$a\cos(ax)$
$\cos(ax)$	$-a\sin(ax)$
$\tan(ax)$	$a(1 + \tan^2(ax))$
$\csc(ax)$	$-a\csc(ax) \cdot \cot(ax)$
$\sec(ax)$	$a\sec(ax) \cdot \tan(ax)$
$\cot(ax)$	$-a(1 + \cot^2(ax))$

Table 4.5 The rules for differentiating inverse trigonometric functions

y	dy/dx
$\arcsin x$	$\dfrac{1}{\sqrt{1-x^2}}$
$\arccos x$	$-\dfrac{1}{\sqrt{1-x^2}}$
$\arctan x$	$\dfrac{1}{1+x^2}$
$\mathrm{arccsc}\, x$	$-\dfrac{1}{x\sqrt{x^2-1}}$
$\mathrm{arcsec}\, x$	$\dfrac{1}{x\sqrt{x^2-1}}$
$\mathrm{arccot}\, x$	$-\dfrac{1}{x^2+1}$

Table 4.6 The rules for integrating trigonometric functions

$f(x)$	$\int f(x)\, dx$
$\sin(ax)$	$-\frac{1}{a}\cos(ax) + C$
$\cos(ax)$	$\frac{1}{a}\sin(ax) + C$
$\sec^2(ax)$	$\frac{1}{a}\tan(ax) + C$
$\csc(ax) \cdot \cot(ax)$	$-\frac{1}{a}\csc(ax) + C$
$\sec(ax) \cdot \tan(ax)$	$\frac{1}{a}\sec(ax) + C$
$\csc^2(ax)$	$-\frac{1}{a}\cot(ax) + C$

Table 4.7 The rules for integrating inverse trigonometric functions

$f(x)$	$\int f(x)\, dx$
$\dfrac{1}{\sqrt{1-x^2}}$	$\arcsin x + C$
$\dfrac{1}{1+x^2}$	$\arctan x + C$
$\dfrac{1}{x\sqrt{x^2-1}}$	$\mathrm{arcsec}\,|x| + C$

4.6 Differentiating Hyperbolic Functions

Trigonometric functions are useful for parametric, circular motion, whereas, hyperbolic functions arise in equations for the absorption of light, mechanics and in integral Calculus. Figure 4.14 shows graphs of the unit circle and a hyperbola whose respective equations are

$$x^2 + y^2 = 1$$
$$x^2 - y^2 = 1$$

where the only difference between them is a sign. The parametric form for the trigonometric, or circular functions and the hyperbolic functions are respectively:

$$\sin^2 \theta + \cos^2 \theta = 1$$
$$\cosh^2 x - \sinh^2 x = 1.$$

The three hyperbolic functions have the following definitions:

$$\sinh x = \frac{e^x - e^{-x}}{2}$$

$$\cosh x = \frac{e^x + e^{-x}}{2}$$

$$\tanh x = \frac{\sinh x}{\cosh x} = \frac{e^{2x} - 1}{e^{2x} + 1}$$

and their reciprocals are:

$$\operatorname{cosech} x = \frac{1}{\sinh x} = \frac{2}{e^x - e^{-x}}$$

$$\operatorname{sech} x = \frac{1}{\cosh x} = \frac{2}{e^x + e^{-x}}$$

$$\coth x = \frac{1}{\tanh x} = \frac{e^{2x} + 1}{e^{2x} - 1}.$$

Other useful identities include:

$$\operatorname{sech}^2 x = 1 - \tanh^2 x$$
$$\operatorname{cosech}^2 = \coth^2 x - 1.$$

The coordinates of P and Q in Fig. 4.14 are given by $P(\cos \theta, \sin \theta)$ and $Q(\cosh x, \sinh x)$. Table 4.8 shows the names of the three hyperbolic functions, their reciprocals and inverse forms. As these functions are based upon e^x and e^{-x}, they are relatively easy to differentiate, which we now investigate.

Fig. 4.14 Graphs of the unit circle $x^2 + y^2 = 1$ and the hyperbola $x^2 - y^2 = 1$

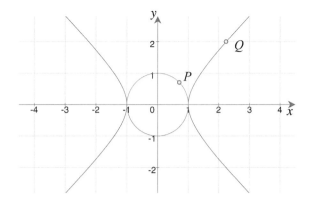

Table 4.8 Hyperbolic function names

Function	Reciprocal	Inverse function	Inverse reciprocal
sinh	cosech	arsinh	arcsch
cosh	sech	arcosh	arsech
tanh	coth	artanh	arcoth

4.6.1 Differentiating sinh, cosh and tanh

The hyperbolic functions are differentiated as follows.
    Let

$$y = \sinh x$$

then

$$y = \frac{e^x - e^{-x}}{2}$$

$$\frac{dy}{dx} = \frac{e^x + e^{-x}}{2}$$

$$\frac{d}{dx} \sinh x = \cosh x.$$

Figure 4.15 shows a graph of $\sinh x$ and its derivative $\cosh x$.
    It follows that

$$\int \cosh x \, dx = \sinh x + C.$$

    Let

$$y = \cosh x$$

Fig. 4.15 Graph of sinh x
and its derivative cosh x

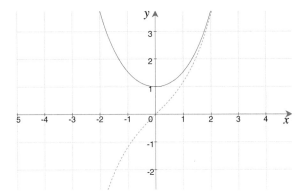

Fig. 4.16 Graph of cosh x
and its derivative sinh x

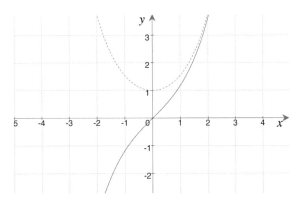

then

$$y = \frac{e^x + e^{-x}}{2}$$

$$\frac{dy}{dx} = \frac{e^x - e^{-x}}{2}$$

$$\frac{d}{dx} \cosh x = \sinh x.$$

Figure 4.16 shows a graph of cosh x and its derivative sinh x.

It follows that

$$\int \sinh x \, dx = \cosh x + C.$$

To differentiate tanh x we employ the quotient rule, and the parametric form of the hyperbola.

Let

$$y = \tanh x$$

Fig. 4.17 Graph of $\tanh x$ and its derivative $\operatorname{sech}^2 x$

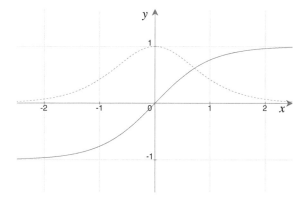

then

$$y = \frac{\sinh x}{\cosh x}$$

$$\frac{dy}{dx} = \frac{\cosh x \cdot \cosh x - \sinh x \cdot \sinh x}{\cosh^2 x}$$

$$= \frac{\cosh^2 x - \sinh^2 x}{\cosh^2 x} = \frac{1}{\cosh^2 x}$$

$$\frac{d}{dx} \tanh x = \operatorname{sech}^2 x.$$

Figure 4.17 shows a graph of $\tanh x$ and its derivative $\operatorname{sech}^2 x$.

4.6.2 *Differentiating cosech, sech and coth*

The hyperbolic reciprocals are differentiated as follows.
  Let

$$y = \operatorname{cosech} x$$

then

$$y = \frac{1}{\sinh x}$$

$$\frac{dy}{dx} = \frac{-\cosh x}{\sinh^2 x}$$

$$\frac{d}{dx} \operatorname{cosech} x = -\operatorname{cosech} x \cdot \coth x.$$

Let

$$y = \text{sech } x$$

then

$$y = \frac{1}{\cosh x}$$

$$\frac{dy}{dx} = \frac{-\sinh x}{\cosh^2 x}$$

$$\frac{d}{dx}\text{sech } x = -\text{sech } x \cdot \tanh x.$$

Let

$$y = \coth x$$

then

$$y = \frac{1}{\tanh x} = \frac{\cosh x}{\sinh x}$$

$$\frac{dy}{dx} = \frac{\sinh^2 x - \cosh^2 x}{\sinh^2 x} = \frac{-1}{\sinh^2 x}$$

$$\frac{d}{dx}\coth x = -\text{cosech}^2 x.$$

4.6.3   Differentiating arsinh, arcosh and artanh

The inverse hyperbolic functions are differentiated as follows.
   Let

$$y = \text{arsinh } x$$

then

$$x = \sinh y$$

$$\frac{dx}{dy} = \cosh y$$

$$\frac{dy}{dx} = \frac{1}{\cosh y} = \frac{1}{\sqrt{1 + \sinh^2 y}}$$

$$\frac{d}{dx}\text{arsinh } x = \frac{1}{\sqrt{1 + x^2}}.$$

It follows that

$$\int \frac{dx}{\sqrt{1+x^2}} = \text{arsinh } x + C.$$

Let

$$y = \text{arcosh } x$$

then

$$x = \cosh y$$

$$\frac{dx}{dy} = \sinh y$$

$$\frac{dy}{dx} = \frac{1}{\sinh y} = \frac{1}{\sqrt{\cosh^2 y - 1}}$$

$$\frac{d}{dx}\text{arcosh } x = \frac{1}{\sqrt{x^2 - 1}}.$$

It follows that

$$\int \frac{dx}{\sqrt{x^2 - 1}} = \text{arcosh } x + C.$$

Let

$$y = \text{artanh } x$$

then

$$x = \tanh y$$

$$\frac{dx}{dy} = \text{sech}^2 y$$

$$\frac{dy}{dx} = \frac{1}{\text{sech}^2 y} = \frac{1}{1 - \tanh^2 y}$$

$$\frac{d}{dx}\text{artanh } x = \frac{1}{1 - x^2}.$$

It follows that

$$\int \frac{dx}{1-x^2} = \text{artanh } x + C.$$

4.6.4   Differentiating arcsch, arsech and arcoth

The inverse, reciprocal hyperbolic functions are differentiated as follows.

Let
$$y = \text{arcsch}\, x$$

then
$$x = \text{cosech}\, y = \frac{1}{\sinh y}$$

$$\frac{dx}{dy} = \frac{-\cosh y}{\sinh^2 y}$$

$$\frac{dy}{dx} = \frac{-\sinh^2 y}{\cosh y}$$

$$\frac{d}{dx}\text{arcsch}\, x = -\frac{1}{x\sqrt{1 + x^2}}.$$

It follows that
$$\int \frac{dx}{x\sqrt{1 + x^2}} = -\text{arcsch}\, x + C.$$

Let
$$y = \text{arsech}\, x$$

then
$$x = \text{sech}\, y = \frac{1}{\cosh y}$$

$$\frac{dx}{dy} = \frac{-\sinh y}{\cosh^2 y}$$

$$\frac{dy}{dx} = \frac{-\cosh^2 y}{\sinh y}$$

$$\frac{d}{dx}\text{arsech}\, x = -\frac{1}{x\sqrt{1 - x^2}}.$$

It follows that
$$\int \frac{dx}{x\sqrt{1 - x^2}} = -\text{arsech}\, x + C.$$

Let
$$y = \text{arcoth}\, x$$

Table 4.9 The rules for differentiating hyperbolic functions

y	dy/dx
$\sinh x$	$\cosh x$
$\cosh x$	$\sinh x$
$\tanh x$	$\text{sech}^2 x$
$\text{cosech}\, x$	$-\text{cosech}\, x \cdot \coth x$
$\text{sech}\, x$	$-\text{sech}\, x \cdot \tanh x$
$\coth x$	$-\text{cosech}^2 x$

Table 4.10 The rules for differentiating inverse hyperbolic functions

y	dy/dx
$\text{arsinh}\, x$	$\dfrac{1}{\sqrt{1+x^2}}$
$\text{arcosh}\, x$	$\dfrac{1}{\sqrt{x^2-1}}$
$\text{artanh}\, x$	$\dfrac{1}{1-x^2}$
$\text{arcsch}\, x$	$-\dfrac{1}{x\sqrt{1+x^2}}$
$\text{arsech}\, x$	$-\dfrac{1}{x\sqrt{1-x^2}}$
$\text{arcoth}\, x$	$-\dfrac{1}{x^2-1}$

then

$$x = \coth y = \frac{\cosh y}{\sinh y}$$

$$\frac{dx}{dy} = \frac{\sinh^2 y - \cosh^2 y}{\sinh^2 y}$$

$$\frac{dy}{dx} = \frac{\sinh^2 y}{\sinh^2 y - \cosh^2 y}$$

$$\frac{d}{dx}\text{arcoth}\, x = -\frac{1}{x^2-1}.$$

It follows that

$$\int \frac{dx}{x^2-1} = -\text{arcoth}\, x + C.$$

Table 4.11 The rules for integrating hyperbolic functions

$f(x)$	$\int f(x)\, dx$
$\sinh x$	$\cosh x + C$
$\cosh x$	$\sinh x + C$
$\operatorname{sech}^2 x$	$\tanh x + C$

Table 4.12 The rules for integrating inverse hyperbolic functions

$f(x)$	$\int f(x)\, dx$
$\dfrac{1}{\sqrt{1+x^2}}$	$\operatorname{arsinh} x + C$
$\dfrac{1}{\sqrt{x^2-1}}$	$\operatorname{arcosh} x + C$
$\dfrac{1}{1-x^2}$	$\operatorname{artanh} x + C$

4.6.5   Summary: Hyperbolic Functions

Table 4.9 shows the rules for differentiating hyperbolic functions, and Table 4.10 shows the rules for the inverse, hyperbolic functions.

Table 4.11 shows the rules for integrating hyperbolic functions, and Table 4.12 shows the rules for integrating inverse, hyperbolic functions.

4.7   Summary

In this chapter we have seen how to differentiate generic functions such as sums, products, quotients and a function of a function, and we have also seen how to address explicit and implicit forms. These techniques were then used to differentiate exponential, logarithmic, trigonometric and hyperbolic functions, which will be employed in later chapters to solve various problems. Where relevant, integrals of certain functions have been included to show the intimate relationship between derivatives and antiderivatives.

Hopefully, it is now clear that differentiation is like an operator—in that it describes how fast a function changes relative to its independent variable in the form of another function. What we have not yet considered is repeated differentiation and its physical meaning, which is the subject of the next chapter.

Chapter 5
Higher Derivatives

5.1 Introduction

There are three sections to this chapter: The first shows what happens when a function is repeatedly differentiated; the second shows how these higher derivatives resolve local minimum and maximum conditions; and the third section provides a physical interpretation for these derivatives. Let's begin by finding the higher derivatives of simple polynomials.

5.2 Higher Derivatives of a Polynomial

We have previously seen that polynomials of the form

$$y = a_n x^n + a_{n-1} x^{n-1} + \cdots + a_2 x^2 + a_1 x + a_0$$

are differentiated as follows:

$$\frac{dy}{dx} = n a_n x^{n-1} + (n-1) a_{n-1} x^{n-2} + \cdots + 2a_2 x + a_1.$$

For example, let

$$y = 3x^3 + 2x^2 - 5x$$

then

$$\frac{dy}{dx} = 9x^2 + 4x - 5$$

which describes how the slope of the original function changes with x.

Figure 5.1 shows the graph of $y = 3x^3 + 2x^2 - 5x$ and its derivative $y = 9x^2 + 4x - 5$, and we can see that when $x = -1$ there is a local maximum, where the

© Springer Nature Switzerland AG 2019
J. Vince, *Calculus for Computer Graphics*,
https://doi.org/10.1007/978-3-030-11376-6_5

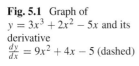

Fig. 5.1 Graph of $y = 3x^3 + 2x^2 - 5x$ and its derivative $\frac{dy}{dx} = 9x^2 + 4x - 5$ (dashed)

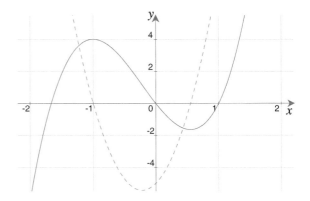

function reaches a value of 4, then begins a downward journey to 0, where the slope is -5. Similarly, when $x \simeq 0.55$, there is a point where the function reaches a local minimum with a value of approximately -1.65. The slope is zero at both points, which is reflected in the graph of the derivative.

Having differentiated the function once, there is nothing to prevent us differentiating a second time, but first we require a way to annotate the process, which is performed as follows. At a general level, let y be some function of x, then the first derivative is

$$\frac{dy}{dx}.$$

The second derivative is found by differentiating the first derivative:

$$\frac{d}{dx}\left(\frac{dy}{dx}\right)$$

and is written:

$$\frac{d^2y}{dx^2}.$$

Similarly, the third derivative is

$$\frac{d^3y}{dx^3}$$

and the nth derivative:

$$\frac{d^ny}{dx^n}.$$

When a function is expressed as $f(x)$, its derivative is written $f'(x)$. The second derivative is written $f''(x)$, and so on for higher derivatives.

Returning to the original function, the first and second derivatives are

$$\frac{dy}{dx} = 9x^2 + 4x - 5$$

$$\frac{d^2y}{dx^2} = 18x + 4$$

and the third and fourth derivatives are

$$\frac{d^3y}{dx^3} = 18$$

$$\frac{d^4y}{dx^4} = 0.$$

Figure 5.2 shows the original function and the first two derivatives. The graph of the first derivative shows the slope of the original function, whereas the graph of the second derivative shows the slope of the first derivative. These graphs help us identify a local maximum and minimum. By inspection of Fig. 5.2, when the first derivative equals zero, there is a local maximum or a local minimum. Algebraically, this is when

$$\frac{dy}{dx} = 0$$

$$9x^2 + 4x - 5 = 0.$$

Solving this quadratic in x we have

$$x = \frac{-b \pm \sqrt{b^2 - 4ac}}{2a}$$

where $a = 9, \quad b = 4, \quad c = -5$:

$$x = \frac{-4 \pm \sqrt{16 + 180}}{18}$$

$$x_1 = -1, \quad x_2 = 0.555$$

which confirms our earlier analysis. However, what we don't know, without referring to the graphs, whether it is a minimum, or a maximum.

5.3 Identifying a Local Maximum or Minimum

Figure 5.3 shows a function containing a local maximum of 5 when $x = -1$. Note that as the independent variable x, increases from -2 towards 0, the slope of the graph changes from positive to negative, passing through zero at $x = -1$. This is

Fig. 5.2 Graph of
$y = 3x^3 + 2x^2 - 5x$, its first
derivative
$\frac{dy}{dx} = 9x^2 + 4x - 5$ (short
dashes) and its second

derivative $\frac{d^2y}{dx^2} = 18x + 4$
(long dashes)

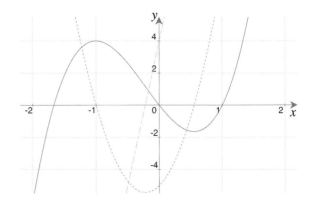

Fig. 5.3 A function
containing a local maximum,
and its first derivative
(dashed)

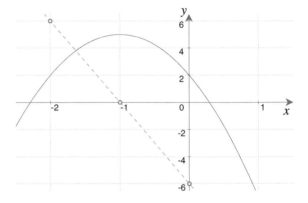

shown in the function's first derivative, which is the straight line passing through the
points $(-2, 6)$, $(-1, 0)$ and $(0, -6)$. In this example these conditions imply that the
slope of the second derivative must be negative:

$$\frac{d^2y}{dx^2} = -\text{ve.}$$

Figure 5.4 shows another function containing a local minimum of 5 when $x = -1$.
Note that as the independent variable x, increases from -2 towards 0, the slope of
the graph changes from negative to positive, passing through zero at $x = -1$. This
is shown in the function's first derivative, which is the straight line passing through
the points $(-2, -6)$, $(-1, 0)$ and $(0, 6)$. In this example these conditions imply that
the slope of the second derivative must be positive:

$$\frac{d^2y}{dx^2} = +\text{ve.}$$

Fig. 5.4 A function containing a local minimum, and its first derivative (dashed)

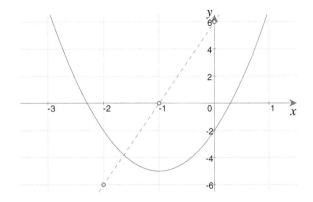

We can now apply this observation to the original function for the two values of x, $x_1 = -1$, $x_2 = 0.555$:

$$\frac{dy}{dx} = 9x^2 + 4x - 5$$

$$\frac{d^2y}{dx^2} = 18x + 4$$

$$= 18 \times (-1) = -18$$

$$= 18 \times (0.555) = +10.$$

Which confirms that when $x = -1$ there is a local maximum, and when $x = 0.555$, there is a local minimum, as shown in Fig. 5.1.

The second derivative test says that **if** the second derivative is positive, evaluated at $x = a$, the solution of

$$\frac{dy}{dx} = 0,$$

then $x = a$ gives a local minimum. Correspondingly, **if** the second derivative is negative, evaluated at $x = a$, the solution of

$$\frac{dy}{dx} = 0,$$

then $x = a$ gives a local maximum.

Let's repeat this technique for

$$y = -3x^3 + 9x$$

whose derivative is

$$\frac{dy}{dx} = -9x^2 + 9$$

Fig. 5.5 Graph of
$y = -3x^3 + 9x$, its first
derivative $y = -9x^2 + 9$
(short dashes) and its second
derivative $y = -18x$ (long
dashes)

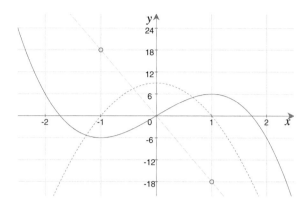

and second derivative

$$\frac{d^2y}{dx^2} = -18x$$

as shown in Fig. 5.5. For a local maximum or minimum, the first derivative equals
zero:

$$-9x^2 + 9 = 0$$

which implies that $x = \pm 1$.

The sign of the second derivative determines whether there is a local minimum
or maximum.

$$\frac{d^2y}{dx^2} = -18x$$
$$= -18 \times (-1) = +ve$$
$$= -18 \times (+1) = -ve$$

therefore, when $x = -1$ there is a local minimum, and when $x = +1$ there is a local
maximum, as confirmed by Fig. 5.5.

5.4   Derivatives and Motion

The first derivative of a simple function of x is its instantaneous slope, which may
be a linear function or some other function. Higher derivatives are the slopes of their
respective functions. For example, for the sine function

$$y = \sin x$$
$$\frac{dy}{dx} = \cos x$$

Table 5.1 The height of an object and distance travelled at different times during its fall

t	d	$s(t)$
0	0	20
0.5	-1.225	18.775
1	-4.9	15.1
1.5	-11.025	8.975
2.02	-20	0

$$\frac{d^2 y}{dx^2} = -\sin x$$

$$\frac{d^3 y}{dx^3} = -\cos x$$

$$\frac{d^4 y}{dx^4} = \sin x.$$

A similar cycle emerges for the cosine function. However, when the independent variable is time, higher derivatives can give the velocity and acceleration of an object, where velocity is the rate of change of position with respect to time, and acceleration is the rate of change of velocity with respect to time.

Let

$$\text{position} = s(t)$$

then

$$\text{velocity } v = \frac{ds}{dt}$$

and

$$\text{acceleration } a = \frac{dv}{dt} = \frac{d^2 s}{dt^2}.$$

For example, when an object is dropped from a height h_0 close to the earth, it experiences a downward acceleration of $g = 9.8$ m/s^2, and falls a distance d:

$$d = -\tfrac{1}{2} g t^2.$$

Observe that a distance measured vertically upwards is positive, and a distance measured downwards is negative. Consequently, its instantaneous height is given by

$$s(t) = h_0 - \tfrac{1}{2} g t^2. \tag{5.1}$$

Figure 5.6 shows the height of the object at different times during its fall, and Table 5.1 gives corresponding values of t, d and $s(t)$, with a starting height $h_0 = 20$ m.

Differentiating (5.1) with respect to time gives the object's instantaneous velocity v:

Fig. 5.6 The position of an
object falling under the pull
of gravity

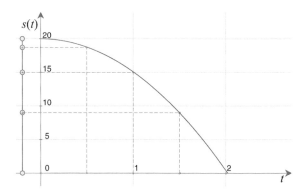

$$s(t) = h_0 - \tfrac{1}{2}gt^2$$

$$v = \frac{ds}{dt} = -gt \tag{5.2}$$

and after 2.02 s, the object is travelling at approximately 19.8 m/s.

Differentiating (5.2) with respect to time gives the instantaneous acceleration of
the object:

$$v = -gt$$

$$a = \frac{dv}{dt} = \frac{d^2s}{dt^2} = -g$$

and after 2.02 s, the object remains accelerating at a constant -9.8 m/s^2.

If the object is subjected to an initial vertical velocity of v_0, after t seconds it
travels a distance of $v_0 t$, which permits us to write a general equation for the object's
height as

$$s(t) = h_0 + v_0 t - \tfrac{1}{2}gt^2. \tag{5.3}$$

Differentiating (5.3) gives the instantaneous velocity:

$$v = \frac{ds}{dt} = v_0 - gt. \tag{5.4}$$

Differentiating (5.4) gives the instantaneous acceleration:

$$a = \frac{dv}{dt} = \frac{d^2s}{dt^2} = -g.$$

If we set the initial velocity to $v_0 = 6$ m/s and maintain the same starting height
$h_0 = 20$, Fig. 5.7 shows the resulting motion.

Fig. 5.7 The position of an object falling under the pull of gravity with an initial upward velocity of 6 m/s

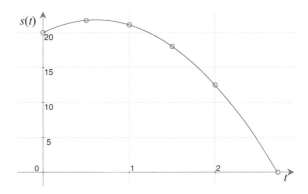

5.5 Summary

In this chapter we have seen how a function can be repeatedly differentiated to reveal higher derivatives. These, in turn, can be used to identify points of local maxima and minima. They can also be used to identify the velocity and acceleration of an object.

5.5.1 Summary of Formulae

Position, Velocity and Acceleration

$$\text{position} = s(t)$$

$$\text{velocity } v = \frac{ds}{dt}$$

$$\text{acceleration } a = \frac{d^2s}{dt^2}.$$

Distance an Object Falls Under Gravity

$$d = -\tfrac{1}{2}gt^2.$$

Instantaneous Height

$$s(t) = h_0 - \tfrac{1}{2}gt^2.$$

Chapter 6
Partial Derivatives

6.1 Introduction

In this chapter we investigate derivatives of functions with more than one independent variable, and how such derivatives are annotated. We also explore the second-order form of these derivatives.

6.2 Partial Derivatives

Up to this point, we have used functions with one independent variable, such as $y = f(x)$. However, we must be able to compute derivatives of functions with more than one independent variable, such as $y = f(u, v, w)$. The technique employed is to assume that only one variable changes, whilst the other variables are held constant. This means that a function can possess several derivatives – one for each independent variable. Such derivatives are called *partial derivatives* and employ a new symbol ∂, which can be read as '*partial dee*'.

Given a function $f(u, v, w)$, the three partial derivatives are defined as

$$\frac{\partial f}{\partial u} = \lim_{h \to 0} \frac{f(u + h, v, w) - f(u, v, w)}{h}$$

$$\frac{\partial f}{\partial v} = \lim_{h \to 0} \frac{f(u, v + h, w) - f(u, v, w)}{h}$$

$$\frac{\partial f}{\partial w} = \lim_{h \to 0} \frac{f(u, v, w + h) - f(u, v, w)}{h}.$$

For example, a function for the volume of a cylinder is

$$V(r, h) = \pi r^2 h$$

© Springer Nature Switzerland AG 2019
J. Vince, *Calculus for Computer Graphics*,
https://doi.org/10.1007/978-3-030-11376-6_6

where r is the radius, and h is the height. Say we wish to compute the function's partial derivative with respect to r. First, the partial derivative is written

$$\frac{\partial V}{\partial r}.$$

Second, we hold h constant, whilst allowing r to change. This means that the function becomes

$$V(r, h) = kr^2 \tag{6.1}$$

where $k = \pi h$. Thus the partial derivative of (6.1) with respect to r is

$$\frac{\partial V}{\partial r} = 2kr$$
$$= 2\pi hr.$$

Next, by holding r constant, and allowing h to change, we have

$$\frac{\partial V}{\partial h} = \pi r^2.$$

Sometimes, for purposes of clarification, the partial derivatives identify the constant variable(s):

$$\left(\frac{\partial V}{\partial r}\right)_h = 2\pi hr$$
$$\left(\frac{\partial V}{\partial h}\right)_r = \pi r^2.$$

Partial differentiation is subject to the same rules for ordinary differentiation – we just to have to remember which independent variable changes, and those held constant. As with ordinary derivatives, we can compute higher-order partial derivatives. For example, consider the function

$$f(u, v) = u^4 + 2u^3v^2 - 4v^3.$$

The first partial derivatives are

$$\frac{\partial f}{\partial u} = 4u^3 + 6u^2v^2$$
$$\frac{\partial f}{\partial v} = 4u^3v - 12v^2$$

and the second-order partial derivatives are

$$\frac{\partial^2 f}{\partial u^2} = 12u^2 + 12uv^2$$

$$\frac{\partial^2 f}{\partial v^2} = 4u^3 - 24v.$$

Similarly, given

$$f(u, v) = \sin(4u) \cdot \cos(5v)$$

the first partial derivatives are

$$\frac{\partial f}{\partial u} = 4\cos(4u) \cdot \cos(5v)$$

$$\frac{\partial f}{\partial v} = -5\sin(4u) \cdot \sin(5v)$$

and the second-order partial derivatives are

$$\frac{\partial^2 f}{\partial u^2} = -16\sin(4u) \cdot \cos(5v)$$

$$\frac{\partial^2 f}{\partial v^2} = -25\sin(4u) \cdot \cos(5v).$$

In general, given $f(u, v) = uv$, then

$$\frac{\partial f}{\partial u} = v$$

$$\frac{\partial f}{\partial v} = u$$

and the second-order partial derivatives are

$$\frac{\partial^2 f}{\partial u^2} = 0$$

$$\frac{\partial^2 f}{\partial v^2} = 0.$$

Similarly, given $f(u, v) = u/v$, then

$$\frac{\partial f}{\partial u} = \frac{1}{v}$$

$$\frac{\partial f}{\partial v} = -\frac{u}{v^2}$$

and the second-order partial derivatives are

$$\frac{\partial^2 f}{\partial u^2} = 0$$

$$\frac{\partial^2 f}{\partial v^2} = \frac{2u}{v^3}.$$

Finally, given $f(u, v) = u^v$, then

$$\frac{\partial f}{\partial u} = vu^{v-1}$$

whereas, $\partial f/\partial v$ requires some explaining. First, given

$$f(u, v) = u^v$$

taking natural logs of both sides, we have

$$\ln f(u, v) = v \ln u$$

and

$$f(u, v) = e^{v \ln u}.$$

Therefore,

$$\frac{\partial f}{\partial v} = e^{v \ln u} \ln u$$

$$= u^v \ln u.$$

The second-order partial derivatives are

$$\frac{\partial^2 f}{\partial u^2} = v(v-1)u^{v-2}$$

$$\frac{\partial^2 f}{\partial v^2} = u^v \ln^2 u.$$

6.2.1   Visualising Partial Derivatives

Functions of the form $y = f(x)$ are represented by a 2D graph, and the function's derivative $f'(x)$ represents the graph's slope at any point x. Functions of the form $z = f(x, y)$ can be represented by a 3D surface, like the one shown in Fig. 6.1, which is $z(x, y) = 4x^2 - 2y^2$. The two partial derivatives are

Fig. 6.1 Surface of
$z = 4x^2 - 2y^2$ using a
right-handed axial system
with a vertical z-axis

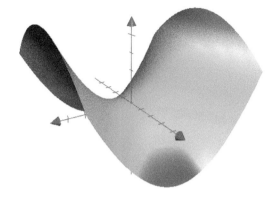

Fig. 6.2 $\partial z/\partial x$ describes the
slopes of these contour lines

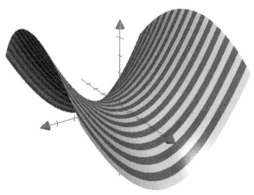

$$\frac{\partial z}{\partial x} = 8x$$

$$\frac{\partial z}{\partial y} = -4y$$

where $\partial z/\partial x$ is the slope of the surface in the x-direction, as shown in Fig. 6.2, and
$\partial z/\partial y$ is the slope of the surface in the y-direction, as shown in Fig. 6.3.

The second-order partial derivatives are

$$\frac{\partial^2 z}{\partial x^2} = 8 = +\text{ve}$$

$$\frac{\partial^2 z}{\partial y^2} = -4 = -\text{ve}.$$

As $\partial^2 z/\partial x^2$ is positive, there is a local minimum in the x-direction, and as $\partial^2 z/\partial y^2$
is negative, there is a local maximum in the y-direction, as confirmed by Figs. 6.2
and 6.3.

Fig. 6.3 $\partial z/\partial y$ describes the slopes of these contour lines

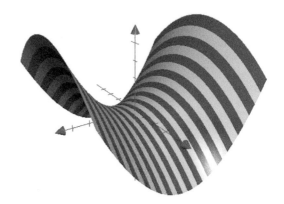

6.2.2 Mixed Partial Derivatives

We have seen that, given a function of the form $f(u, v)$, the partial derivatives $\partial f/\partial u$ and $\partial f/\partial v$ provide the relative instantaneous changes in f and u, and f and v, respectively, whilst the second independent variable remains fixed. However, nothing prevents us from differentiating $\partial f/\partial u$ with respect to v, whilst keeping u constant:

$$\frac{\partial}{\partial v}\left(\frac{\partial f}{\partial u}\right)$$

which is also written as

$$\frac{\partial^2 f}{\partial v \partial u}$$

and is a *mixed partial derivative*. For example, let

$$f(u, v) = u^3 v^4$$

then

$$\frac{\partial f}{\partial u} = 3u^2 v^4$$

and

$$\frac{\partial^2 f}{\partial v \partial u} = 12u^2 v^3.$$

However, it should be no surprise that reversing the differentiation gives the same result. Let

$$f(u, v) = u^3 v^4$$

then

$$\frac{\partial f}{\partial v} = 4u^3 v^3$$

and

$$\frac{\partial^2 f}{\partial u \partial v} = 12u^2 v^3.$$

Generally, for continuous functions, we can write

$$\frac{\partial^2 f}{\partial u \partial v} = \frac{\partial^2 f}{\partial v \partial u}.$$

For example, the formula for the volume of a cylinder is given by $V(r, h) = \pi r^2 h$, where r and h are the cylinder's radius and height, respectively. The mixed partial derivative is computed as follows.

$$V(r, h) = \pi r^2 h$$
$$\frac{\partial V}{\partial r} = 2\pi h r$$
$$\frac{\partial^2 V}{\partial h \partial r} = 2\pi r$$

or

$$V(r, h) = \pi r^2 h$$
$$\frac{\partial V}{\partial h} = \pi r^2$$
$$\frac{\partial^2 V}{\partial r \partial h} = 2\pi r.$$

As a second example, let

$$f(u, v) = \sin(4u) \cdot \cos(3v)$$

then

$$\frac{\partial f}{\partial u} = 4\cos(4u) \cdot \cos(3v)$$
$$\frac{\partial^2 f}{\partial v \partial u} = -12\cos(4u) \cdot \sin(3v)$$

or

$$\frac{\partial f}{\partial v} = -3\sin(4u) \cdot \sin(3v)$$
$$\frac{\partial^2 f}{\partial u \partial v} = -12\cos(4u) \cdot \sin(3v).$$

6.3   Chain Rule

In Chap. 4 we came across the chain rule for computing the derivatives of functions of functions. For example, to compute the derivative of $y = \sin(x^2)$ we substitute $u = x^2$, then

$$y = \sin u$$
$$\frac{dy}{du} = \cos u$$
$$= \cos(x^2).$$

Next, we compute du/dx:

$$u = x^2$$
$$\frac{du}{dx} = 2x$$

and dy/dx is the product of the two derivatives using the chain rule:

$$\frac{dy}{dx} = \frac{dy}{du} \cdot \frac{du}{dx}$$
$$= \cos(x^2) \cdot 2x$$
$$= 2x \cos(x^2).$$

But say we have a function where w is a function of two variables x and y, which in turn, are a function of u and v. Then we have

$$w = f(x, y)$$
$$x = r(u, v)$$
$$y = s(u, v).$$

With such a scenario, we have the following partial derivatives:

$$\frac{\partial w}{\partial x}, \quad \frac{\partial w}{\partial y}$$
$$\frac{\partial w}{\partial u}, \quad \frac{\partial w}{\partial v}$$
$$\frac{\partial x}{\partial u}, \quad \frac{\partial x}{\partial v}$$
$$\frac{\partial y}{\partial u}, \quad \frac{\partial y}{\partial v}.$$

These are chained together as follows:

$$\frac{\partial w}{\partial u} = \frac{\partial w}{\partial x} \cdot \frac{\partial x}{\partial u} + \frac{\partial w}{\partial y} \cdot \frac{\partial y}{\partial u} \tag{6.2}$$

$$\frac{\partial w}{\partial v} = \frac{\partial w}{\partial x} \cdot \frac{\partial x}{\partial v} + \frac{\partial w}{\partial y} \cdot \frac{\partial y}{\partial v}. \tag{6.3}$$

For example, given

$$w(x, y) = 2x + 3y$$
$$x(u, v) = u^2 + v^2$$
$$y(u, v) = u^2 - v^2$$

then

$$\frac{\partial w}{\partial x} = 2, \quad \frac{\partial w}{\partial y} = 3$$

$$\frac{\partial x}{\partial u} = 2u, \quad \frac{\partial x}{\partial v} = 2v$$

$$\frac{\partial y}{\partial u} = 2u, \quad \frac{\partial y}{\partial v} = -2v$$

and plugging these into (6.2) and (6.3) we have

$$\frac{\partial w}{\partial u} = \frac{\partial w}{\partial x} \cdot \frac{\partial x}{\partial u} + \frac{\partial w}{\partial y} \cdot \frac{\partial y}{\partial u}$$
$$= 2 \times 2u + 3 \times 2u$$
$$= 10u$$

$$\frac{\partial w}{\partial v} = \frac{\partial w}{\partial x} \cdot \frac{\partial x}{\partial v} + \frac{\partial w}{\partial y} \cdot \frac{\partial y}{\partial v}$$
$$= 2 \times 2v + 3 \times (-2v)$$
$$= -2v.$$

Thus, when $u = 2$ and $v = 1$

$$\frac{\partial w}{\partial u} = 20, \quad \text{and} \quad \frac{\partial w}{\partial v} = -2.$$

6.4   Total Derivative

Given a function with three independent variables, such as $w = f(x, y, t)$, where $x = g(t)$ and $y = h(t)$, there are three primary partial derivatives

$$\frac{\partial w}{\partial x}, \quad \frac{\partial w}{\partial y} \quad \text{and} \quad \frac{\partial w}{\partial t}$$

which show the differential change of w with x, y and t respectively. There are also three derivatives

$$\frac{dx}{dt}, \quad \frac{dy}{dt} \quad \text{and} \quad \frac{dt}{dt}$$

where $dt/dt = 1$. The partial and ordinary derivatives can be combined to create the *total derivative* which is written

$$\frac{dw}{dt} = \frac{\partial w}{\partial x} \cdot \frac{dx}{dt} + \frac{\partial w}{\partial y} \cdot \frac{dy}{dt} + \frac{\partial w}{\partial t}.$$

dw/dt measures the instantaneous change of w relative to t, when all three independent variables change. For example, given

$$w(x, y, t) = x^2 + xy + y^3 + t^2$$
$$x(t) = 2t$$
$$y(t) = t - 1$$

then

$$\frac{dx}{dt} = 2$$

$$\frac{dy}{dt} = 1$$

$$\frac{\partial w}{\partial x} = 2x + y = 4t + t - 1 = 5t - 1$$

$$\frac{\partial w}{\partial y} = x + 3y^2 = 2t + 3(t - 1)^2 = 3t^2 - 4t + 3$$

$$\frac{\partial w}{\partial t} = 2t$$

$$\frac{dw}{dt} = \frac{\partial w}{\partial x} \cdot \frac{dx}{dt} + \frac{\partial w}{\partial y} \cdot \frac{dy}{dt} + \frac{\partial w}{\partial t}$$
$$= (5t - 1)2 + (3t^2 - 4t + 3) + 2t = 3t^2 + 8t + 1$$

and the total derivative equals

$$\frac{dw}{dt} = 3t^2 + 8t + 1$$

and when $t = 1$, $dw/dt = 12$.

6.5 Second-Order and Higher Partial Derivatives

Like ordinary derivatives, it is also possible to take second-order and higher partial derivatives.

6.6 Summary

When a function has two or more independent variables, a partial derivative records the instantaneous rate of change relative to one variable, while the others are held constant.

6.6.1 Summary of Formulae

Mixed Partial Derivatives

$$w = f(u, v)$$

$$\frac{\partial^2 w}{\partial u \partial v} = \frac{\partial^2 w}{\partial v \partial u}.$$

The Chain Rule

$$w = f(x, y)$$
$$x = r(u, v)$$
$$y = s(u, v)$$
$$\frac{\partial w}{\partial u} = \frac{\partial w}{\partial x} \cdot \frac{\partial x}{\partial u} + \frac{\partial w}{\partial y} \cdot \frac{\partial y}{\partial u}$$
$$\frac{\partial w}{\partial v} = \frac{\partial w}{\partial x} \cdot \frac{\partial x}{\partial v} + \frac{\partial w}{\partial y} \cdot \frac{\partial y}{\partial v}.$$

The Total Derivative

$$w = f(x, y, t)$$
$$\frac{dw}{dt} = \frac{\partial w}{\partial x} \cdot \frac{dx}{dt} + \frac{\partial w}{\partial y} \cdot \frac{dy}{dt} + \frac{\partial w}{\partial t}.$$

Chapter 7
Integral Calculus

7.1 Introduction

In this chapter I develop the idea that integration is the inverse of differentiation, and examine standard algebraic strategies for integrating functions, where the derivative is unknown; these include simple algebraic manipulation, trigonometric identities, integration by parts, integration by substitution and integration using partial fractions.

7.2 Indefinite Integral

In previous chapters we have seen that given a simple function, such as

$$y = \sin x + 23$$
$$\frac{dy}{dx} = \cos x$$

and the constant term 23 disappears. Inverting the process, we begin with

$$dy = \cos x \, dx$$

and integrate both sides:

$$y = \int \cos x \, dx$$
$$= \sin x + C.$$

An integral of the form

$$\int f(x) \, dx$$

© Springer Nature Switzerland AG 2019
J. Vince, *Calculus for Computer Graphics*,
https://doi.org/10.1007/978-3-030-11376-6_7

is known as an *indefinite integral*; and as we don't know whether the original function contains a constant term, a constant C has to be included. Its value remains undetermined unless we are told something about the original function. In this example, if we are told that when $x = \frac{\pi}{2}$, $y = 24$, then

$$
\begin{aligned}
24 &= \sin\left(\frac{\pi}{2}\right) + C \\
&= 1 + C \\
C &= 23.
\end{aligned}
$$

7.3   Standard Integration Formulae

In earlier chapters, I have included indefinite integrals for most of the derivatives we have examined. For example, knowing that

$$
\frac{d}{dx} \sin x = \cos x
$$

then the inverse operation is

$$
\int \cos x \, dx = \sin x + C.
$$

For convenience, here they are again:

$$
\int x^n \, dx = \frac{1}{n+1} x^{n+1} + C; \quad n \neq -1
$$

$$
\int e^x \, dx = e^x + C
$$

$$
\int e^{-x} \, dx = -e^{-x} + C
$$

$$
\int e^{ax} \, dx = \frac{1}{a} e^{ax} + C
$$

$$
\int e^{-ax} \, dx = -\frac{1}{a} e^{-ax} + C
$$

$$
\int a^x \, dx = \frac{1}{\ln a} a^x + C; \quad 0 < a \neq 1
$$

$$
\int a^{-x} \, dx = -\frac{1}{\ln a} a^{-x} + C
$$

$$
\int \sin(ax) \, dx = -\frac{1}{a} \cos(ax) + C
$$

$$\int \cos(ax)\, dx = \tfrac{1}{a} \sin(ax) + C$$

$$\int \sec^2(ax)\, dx = \tfrac{1}{a} \tan(ax) + C$$

$$\int \csc(ax) \cdot \cot(ax)\, dx = -\tfrac{1}{a} \csc(ax) + C$$

$$\int \sec(ax) \cdot \tan(ax)\, dx = \tfrac{1}{a} \sec(ax) + C$$

$$\int \csc^2(ax)\, dx = -\tfrac{1}{a} \cot(ax) + C$$

$$\int \frac{1}{\sqrt{1-x^2}}\, dx = \arcsin x + C$$

$$\int \frac{1}{1+x^2}\, dx = \arctan x + C$$

$$\int \frac{1}{x\sqrt{x^2-1}}\, dx = \operatorname{arcsec} |x| + C$$

$$\int \sinh x\, dx = \cosh x + C$$

$$\int \cosh x\, dx = \sinh x + C$$

$$\int \operatorname{sech}^2 x\, dx = \tanh x + C$$

$$\int \frac{1}{\sqrt{1+x^2}}\, dx = \operatorname{arsinh} x + C$$

$$\int \frac{1}{\sqrt{x^2-1}}\, dx = \operatorname{arcosh} x + C$$

$$\int \frac{1}{1-x^2}\, dx = \operatorname{artanh} x + C.$$

All the above integrals, and many more, can be found on the internet and in most books on Calculus. However, the problem facing us now is how to integrate functions that don't fall into the above formats, which is what we consider next.

7.4 Integration Techniques

7.4.1 Continuous Functions

Functions come in all sorts of shapes and sizes, which is why we have to be very careful before they are differentiated or integrated. If a function contains any form of discontinuity, then it cannot be differentiated or integrated. For example, the

Fig. 7.1 A discontinuous
square-wave function

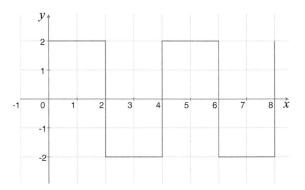

square-wave function shown in Fig. 7.1 cannot be differentiated as it contains discontinuities. Consequently, to be very precise, we identify an *interval* $[a, b]$, over which a function is analysed, and stipulate that it must be continuous over this interval. For example, a and b define the upper and lower bounds of the interval $x \in [a, b]$, then we can say that for $f(x)$ to be continuous

$$\lim_{h \to 0} f(x + h) = f(x).$$

Even this needs further clarification as h must not take x outside of the permitted interval. So, from now on, we assume that all functions are continuous and can be integrated without fear of singularities.

7.4.2   Difficult Functions

Some functions cannot be differentiated easily. For example, the derivative of $\sin x / x$ does not exist, which precludes the possibility of its integration. Figure 7.2 shows

Fig. 7.2 Graph of
$y = (\sin x)/x$

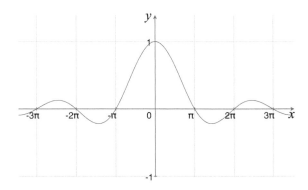

Fig. 7.3 Graph of
$y = \sqrt{x}\sin x$

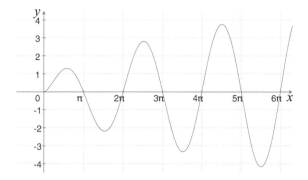

this function, and even though it is continuous, its derivative and integral can only be approximated. Similarly, the derivative of $\sqrt{x}\sin x$ does not exist, and also precludes the possibility of its integration. Figure 7.3 shows this continuous function. So now let's examine how most functions have to be rearranged to secure their integration.

Let's demonstrate through a series of examples how a function can be manipulated to permit it to be integrated.

7.4.3 Trigonometric Identities

Sometimes it is possible to simplify the integrand by substituting a trigonometric identity. For example, the identity $\sin^2 x = \frac{1}{2}(1 - \cos(2x))$ converts the square function of x into a double-angle representation:

$$\int \sin^2 x \, dx = \frac{1}{2}\int (1 - \cos(2x)) \, dx$$
$$= \frac{1}{2}\int 1 \, dx - \frac{1}{2}\int \cos(2x) \, dx$$
$$= \frac{1}{2}x - \frac{1}{4}\sin(2x) + C.$$

Figure 7.4 shows the graphs of $y = \sin^2 x$ and $y = \frac{1}{2}x - \frac{1}{4}\sin(2x)$.

Similarly, the identity $\cos^2 x = \frac{1}{2}(\cos(2x) + 1)$ converts the square function of x into a double-angle representation:

$$\int \cos^2 x \, dx = \frac{1}{2}\int (\cos(2x) + 1) \, dx$$
$$= \frac{1}{2}\int \cos(2x) \, dx + \frac{1}{2}\int 1 \, dx$$
$$= \frac{1}{4}\sin(2x) + \frac{1}{2}x + C.$$

Fig. 7.4 The graphs of
$y = \sin^2 x$ (broken line) and
$y = \frac{1}{2}x - \frac{1}{4}\sin(2x)$

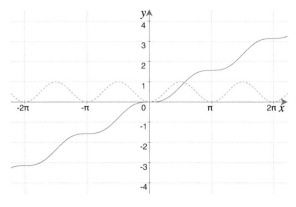

Fig. 7.5 The graphs of
$y = \cos^2 x$ (broken line) and
$y = \frac{1}{4}\sin(2x) + \frac{1}{2}x$

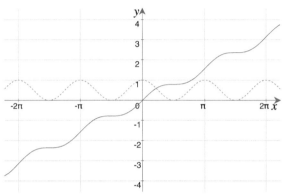

Figure 7.5 shows the graphs of $y = \cos^2 x$ and $y = \frac{1}{4}\sin(2x) + \frac{1}{2}x$.
To integrate $\tan^2 x$ we use the identity $\sec^2 x = 1 + \tan^2 x$:

$$\int \tan^2 x \, dx = \int (\sec^2 x - 1) \, dx$$

$$= \int \sec^2 x \, dx - \int 1 \, dx$$

$$= \tan x - x + C.$$

Figure 7.6 shows the graphs of $y = \tan^2 x$ and $y = \tan x - x$.
To evaluate $\int \sin(3x) \cdot \cos x \, dx$, we use the identity:

$$2 \sin a \cdot \cos b = \sin(a + b) + \sin(a - b)$$

Fig. 7.6 The graphs of
$y = \tan^2 x$ (broken line) and
$y = \tan x - x$

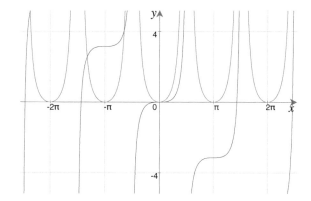

Fig. 7.7 The graphs of
$y = \sin(3x) \cdot \cos x$ (broken
line) and $y =$
$-\frac{1}{8}\cos(4x) - \frac{1}{4}\cos(2x)$

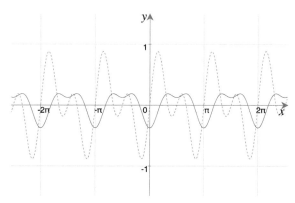

which converts the integrand's product into the sum and difference of two angles:

$$\sin(3x) \cdot \cos x = \tfrac{1}{2}\left(\sin(4x) + \sin(2x)\right)$$

$$\int \sin(3x) \cdot \cos x \, dx = \tfrac{1}{2}\int \sin(4x) + \sin(2x) \, dx$$

$$= \tfrac{1}{2}\int \sin(4x) \, dx + \tfrac{1}{2}\int \sin(2x) \, dx$$

$$= -\tfrac{1}{8}\cos(4x) - \tfrac{1}{4}\cos(2x) + C.$$

Figure 7.7 shows the graphs of $y = \sin(3x) \cdot \cos x$ and $y = -\frac{1}{8}\cos(4x) - \frac{1}{4}\cos(2x)$.

7.4.4 Exponent Notation

Sometimes it's convenient to replace radicals by exponent notation. For example, to
evaluate $\int \frac{2}{\sqrt[3]{x}} \, dx$, the 2 is moved outside the integral, and the integrand is converted

Fig. 7.8 The graphs of
$y = 2/\sqrt[4]{x}$ (broken line) and
$y = 8x^{\frac{3}{4}}/3$

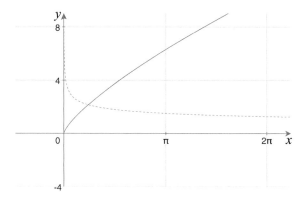

into an exponent form:

$$2 \int \frac{1}{\sqrt[4]{x}} \, dx = 2 \int x^{-\frac{1}{4}}$$

$$= 2 \left(\frac{x^{\frac{3}{4}}}{\frac{3}{4}} \right) + C$$

$$= 2 \left(\frac{4}{3} x^{\frac{3}{4}} \right) + C$$

$$= \frac{8}{3} x^{\frac{3}{4}} + C.$$

Figure 7.8 shows the graphs of $y = 2/\sqrt[4]{x}$ and $y = 8x^{\frac{3}{4}}/3$.

7.4.5   Completing the Square

Sometimes, an integrand can be simplified by completing the square. For example, to evaluate

$$\int \frac{1}{x^2 - 4x + 8} \, dx$$

we note the following.
    We have already seen that

$$\int \frac{1}{1 + x^2} \, dx = \arctan x + C$$

and it's not too difficult to prove that

$$\int \frac{1}{a^2 + x^2} \, dx = \frac{1}{a} \arctan \left(\frac{x}{a} \right) + C.$$

Fig. 7.9 The graphs of
$y = 1/(x^2 - 4x + 8)$
(broken line) and
$y = \frac{1}{2} \arctan \left(\frac{x-2}{2} \right)$

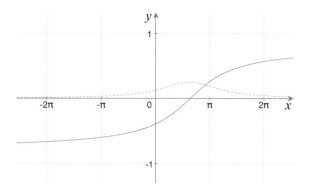

Therefore, if we can manipulate an integrand into this form, then the integral will reduce to an arctan result. The following needs no manipulation:

$$\int \frac{1}{4 + x^2}\, dx = \frac{1}{2} \arctan \left(\frac{x}{2} \right) + C.$$

However, the original integrand has $x^2 - 4x + 8$ as the denominator, which is resolved by completing the square:

$$x^2 - 4x + 8 = 4 + (x - 2)^2.$$

Therefore,

$$\int \frac{1}{x^2 - 4x + 8}\, dx = \int \frac{1}{2^2 + (x - 2)^2}\, dx$$
$$= \frac{1}{2} \arctan \left(\frac{x - 2}{2} \right) + C.$$

Figure 7.9 shows the graphs of $y = 1/(x^2 - 4x + 8)$ and $y = \frac{1}{2} \arctan \left(\frac{x-2}{2} \right)$.
To evaluate

$$\int \frac{1}{x^2 + 6x + 10}\, dx.$$

we use the above arctan function as follows

$$\int \frac{1}{x^2 + 6x + 10}\, dx = \int \frac{1}{1^2 + (x + 3)^2}\, dx$$
$$= \arctan(x + 3) + C.$$

Figure 7.10 shows the graphs of $y = 1/(x^2 + 6x + 10)$ and $y = \arctan(x + 3)$.

Fig. 7.10 The graphs of
$y = 1/(x^2 + 6x + 10)$
(broken line) and
$y = \arctan(x + 3)$

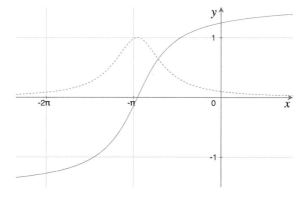

7.4.6   The Integrand Contains a Derivative

Let's consider the situation when the integrand contains a function and its derivative, as in

$$\int \frac{\arctan x}{1 + x^2}\, dx.$$

Knowing that

$$\frac{d}{dx} \arctan x = \frac{1}{1 + x^2}$$

let $u = \arctan x$, then

$$\frac{du}{dx} = \frac{1}{1 + x^2}$$

and

$$\int \frac{\arctan x}{1 + x^2}\, dx = \int u\, du$$
$$= \tfrac{1}{2} u^2 + C$$
$$= \tfrac{1}{2} \arctan^2 x + C.$$

Figure 7.11 shows the graphs of $y = \arctan x/(1 + x^2)$ and $y = \tfrac{1}{2} \arctan^2 x$.
   Here is another example involving $\sin x$ and $\cos x$:

$$\int \frac{\cos x}{\sin x}\, dx.$$

Knowing that

$$\frac{d}{dx} \sin x = \cos x$$

Fig. 7.11 The graphs of
$y = \arctan x/(1 + x^2)$
(broken line) and
$y = \frac{1}{2} \arctan^2 x$

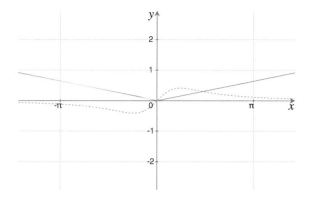

Fig. 7.12 The graphs of
$y = \cos x/\sin x$ (broken
line) and $y = \ln|\sin x|$

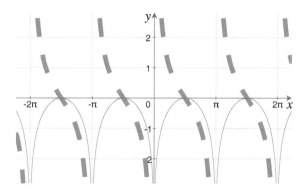

let $u = \sin x$, then

$$\frac{du}{dx} = \cos x$$

and

$$\int \frac{\cos x}{\sin x} \, dx = \int \frac{1}{u} \, du$$
$$= \ln|u| + C$$
$$= \ln|\sin x| + C.$$

Figure 7.12 shows the graphs of $y = \cos x/\sin x$ and $y = \ln|\sin x|$.

Now let's reverse the $\sin x$ and $\cos x$ functions:

$$\int \frac{\sin x}{\cos x} \, dx.$$

Knowing that

$$\frac{d}{dx} \cos x = -\sin x$$

Fig. 7.13 The graphs of
$y = \sin x / \cos x$ (broken
line) and $y = \ln |\sec x|$

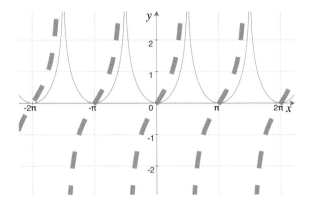

let $u = \cos x$, then

$$\frac{du}{dx} = -\sin x$$

$$du = -\sin x \, dx$$

and

$$\int \frac{\sin x}{\cos x} \, dx = -\int \frac{1}{u} \, du$$
$$= -\ln |u| + C$$
$$= -\ln |\cos x| + C$$
$$= \ln |\cos x|^{-1} + C$$
$$= \ln |\sec x| + C.$$

Figure 7.13 shows the graphs of $y = \sin x / \cos x$ and $y = \ln |\sec x|$.

7.4.7   Converting the Integrand into a Series of Fractions

Integration is often made easier by converting an integrand into a series of fractions.
Here are two examples where the denominator is divided into each term of the
numerator.

$$\int \frac{4x^3 + x^2 - 8 + 12x \cos x}{4x} \, dx.$$

Fig. 7.14 The graphs of
$y = (4x^3 + x^2 - 8 + 12x \cos x)/4x$ (broken line)
and $y = \frac{1}{3}x^3 + \frac{1}{8}x^2 - 2 \ln|x| + 3 \sin x$

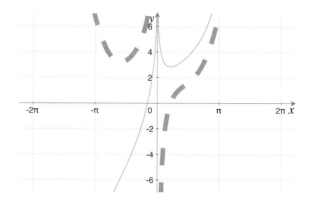

$$\int \frac{4x^3 + x^2 - 8 + 12x \cos x}{4x} \, dx = \int x^2 \, dx + \frac{1}{4} \int x \, dx - 2 \int \frac{1}{x} \, dx + 3 \int \cos x \, dx$$

$$= \frac{1}{3}x^3 + \frac{1}{8}x^2 - 2 \ln|x| + 3 \sin x + C.$$

Figure 7.14 shows the graphs of $y = (4x^3 + x^2 - 8 + 12x \cos x)/4x$ and $y = \frac{1}{3}x^3 + \frac{1}{8}x^2 - 2 \ln|x| + 3 \sin x$.

In this example the denominator $\cos x$ is divided into the three terms of the numerator:

$$\int \frac{2 \sin x + \cos x + \sec x}{\cos x} \, dx.$$

$$\int \frac{2 \sin x + \cos x + \sec x}{\cos x} \, dx = 2 \int \tan x \, dx + \int 1 \, dx + \int \sec^2 x \, dx$$

$$= 2 \ln|\sec x| + x + \tan x + C.$$

Figure 7.15 shows the graphs of $y = (2 \sin x + \cos x + \sec x)/\cos x$ and $y = 2 \ln|\sec x| + x + \tan x$.

7.4.8 Integration by Parts

Integration by parts is based upon the rule for differentiating function products where

$$\frac{d}{dx} uv = u \frac{dv}{dx} + v \frac{du}{dx}$$

Fig. 7.15 The graphs of
$y = (2 \sin x + \cos x + \sec x)/\cos x$ (broken line)
and
$y = 2 \ln|\sec x| + x + \tan x$

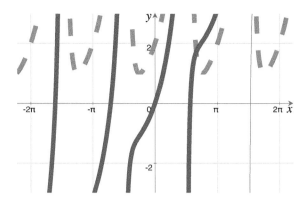

and integrating throughout, we have

$$uv = \int uv' \, dx + \int vu' \, dx$$

which rearranged, gives

$$\int uv' \, dx = uv - \int vu' \, dx.$$

Thus, if an integrand contains a product of two functions, we can attempt to integrate it by parts. Let's start with

$$\int x \sin x \, dx.$$

In this case, we try the following:

$$u = x \quad \text{and} \quad v' = \sin x$$

therefore

$$u' = 1 \quad \text{and} \quad v = C_1 - \cos x.$$

Integrating by parts:

$$\int uv' \, dx = uv - \int vu' \, dx$$
$$\int x \sin x \, dx = x(C_1 - \cos x) - \int (C_1 - \cos x)(1) \, dx$$
$$= C_1 x - x \cos x - C_1 x + \sin x + C$$
$$= -x \cos x + \sin x + C.$$

Fig. 7.16 The graphs of
$y = x \sin x$ (broken line) and
$y = -x \cos x + \sin x$

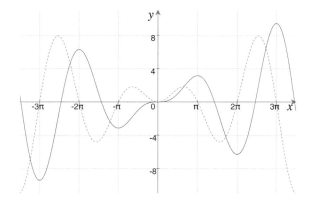

Figure 7.16 shows the graphs of $y = x \sin x$ and $y = -x \cos x + \sin x$.

Note the problems that arise if we make the wrong substitution:

$$u = \sin x \quad \text{and} \quad v' = x$$

therefore

$$u' = \cos x \quad \text{and} \quad v = \tfrac{1}{2}x^2 + C_1$$

Integrating by parts:

$$\int uv' \, dx = uv - \int vu' \, dx$$

$$\int x \sin x \, dx = \sin x \left(\tfrac{1}{2}x^2 + C_1\right) - \int \left(\tfrac{1}{2}x^2 + C_1\right) \cos x \, dx$$

which requires to be integrated by parts, and is even more difficult, which suggests that we made the wrong substitution.

Now let's try something similar, but with the cos function:

$$\int x \cos x \, dx.$$

In this case, we try the following:

$$u = x \quad \text{and} \quad v' = \cos x$$

therefore

$$u' = 1 \quad \text{and} \quad v = \sin x + C_1.$$

Fig. 7.17 The graphs of
$y = x \cos x$ (broken line)
and $y = x \sin x + \cos x$

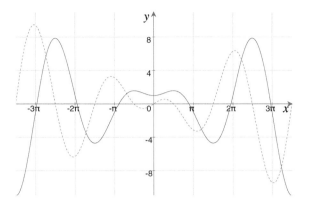

Integrating by parts:

$$\int uv' \, dx = uv - \int vu' \, dx$$

$$\int x \cos x \, dx = x(\sin x + C_1) - \int (\sin x + C_1)(1) \, dx$$

$$= x \sin x + C_1 x + \cos x - C_1 x + C$$

$$= x \sin x + \cos x + C.$$

Figure 7.17 shows the graphs of $y = x \cos x$ and $y = x \sin x + \cos x$.
 Let's develop the last example by changing the x multiplier into x^2:

$$\int x^2 \cos x \, dx.$$

In this case, we try the following:

$$u = x^2 \quad \text{and} \quad v' = \cos x$$

therefore

$$u' = 2x \quad \text{and} \quad v = \sin x + C_1.$$

Integrating by parts:

$$\int uv' \, dx = uv - \int vu' \, dx$$

$$\int x^2 \cos x \, dx = x^2(\sin x + C_1) - 2 \int (\sin x + C_1)(x) \, dx$$

Fig. 7.18 The graphs of
$y = x^2 \cos x$ (broken line)
and $y =$
$x^2 \sin x + 2x \cos x - 2 \sin x$

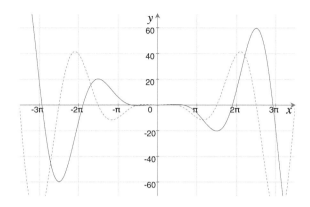

$$= x^2 \sin x + C_1 x^2 - 2C_1 \int x \, dx - 2 \int x \sin x \, dx$$

$$= x^2 \sin x + C_1 x^2 - 2C_1 \left(\tfrac{1}{2} x^2 + C_2 \right) - 2 \int x \sin x \, dx$$

$$= x^2 \sin x - C_3 - 2 \int x \sin x \, dx.$$

At this point we come across $\int x \sin x \, dx$, which we have already solved:

$$\int x^2 \cos x \, dx = x^2 \sin x - C_3 - 2(-x \cos x + \sin x + C_4)$$

$$= x^2 \sin x - C_3 + 2x \cos x - 2 \sin x - C_5$$

$$= x^2 \sin x + 2x \cos x - 2 \sin x + C$$

Figure 7.18 shows the graphs of $y = x^2 \cos x$ and $y = x^2 \sin x + 2x \cos x - 2 \sin x$.
Now let's evaluate

$$\int x^2 \sin x \, dx.$$

In this case, we try the following:

$$u = x^2 \quad \text{and} \quad v' = \sin x$$

therefore

$$u' = 2x \quad \text{and} \quad v = - \cos x + C_1.$$

Fig. 7.19 The graphs of
$y = x^2 \sin x$ (broken line)
and $y = -x^2 \cos x +$
$2x \sin x + 2 \cos x$

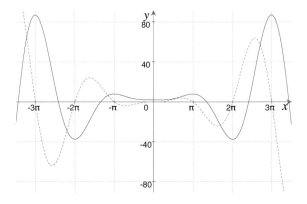

Integrating by parts:

$$\int uv' \, dx = uv - \int vu' \, dx$$

$$\int x^2 \sin x \, dx = x^2(-\cos x + C_1) - 2 \int (-\cos x + C_1)(x) \, dx$$

$$= -x^2 \cos x + C_1 x^2 - 2C_1 \int x \, dx + 2 \int x \cos x \, dx$$

$$= -x^2 \cos x + C_1 x^2 - 2C_1 \left(\tfrac{1}{2}x^2 + C_2\right) + 2 \int x \cos x \, dx$$

$$= -x^2 \cos x - C_3 + 2 \int x \cos x \, dx.$$

At this point we come across $\int x \cos x \, dx$, which we have already solved:

$$\int x^2 \sin x \, dx = -x^2 \cos x - C_3 + 2(x \sin x + \cos x + C_4)$$

$$= -x^2 \cos x - C_3 + 2x \sin x + 2 \cos x + C_5$$

$$= -x^2 \cos x + 2x \sin x + 2 \cos x + C$$

Figure 7.19 shows the graphs of $y = x^2 \sin x$ and $y = -x^2 \cos x + 2x \sin x + 2 \cos x$.

In future, we omit the integration constant, as it is cancelled out during the integration calculation. The next example is

$$\int x \ln x \, dx.$$

Fig. 7.20 The graphs of
$y = x \ln x$ (broken line) and
$y = \frac{1}{2}x^2 \ln x - \frac{1}{4}x^2$

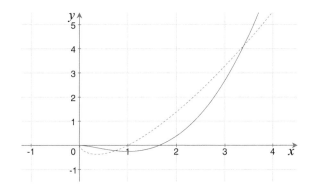

In this case, we try the following:

$$u = \ln x \quad \text{and} \quad v' = x$$

therefore

$$u' = \frac{1}{x} \quad \text{and} \quad v = \frac{1}{2}x^2.$$

Integrating by parts:

$$\int uv' \, dx = uv - \int vu' \, dx$$

$$\int x \ln x \, dx = \frac{1}{2}x^2 \ln x - \int \left(\frac{1}{2}x^2\right)\frac{1}{x} \, dx$$

$$= \frac{1}{2}x^2 \ln x - \frac{1}{2}\int x \, dx$$

$$= \frac{1}{2}x^2 \ln x - \frac{1}{4}x^2 + C.$$

Figure 7.20 shows the graphs of $y = x \ln x$ and $y = \frac{1}{2}x^2 \ln x - \frac{1}{4}x^2$.

Although the following integrand does not look as though it can be integrated by parts,

$$\int \sqrt{1 + x^2} \, dx.$$

if we rewrite it as

$$\int \sqrt{1 + x^2}(1) \, dx.$$

then we can use the formula.

Let

$$u = \sqrt{1 + x^2} \quad \text{and} \quad v' = 1$$

therefore

$$u' = \frac{x}{\sqrt{1+x^2}} \quad \text{and} \quad v = x.$$

Integrating by parts:

$$\int uv' \, dx = uv - \int vu' \, dx$$

$$\int \sqrt{1+x^2} \, dx = x\sqrt{1+x^2} - \int \frac{x^2}{\sqrt{1+x^2}} \, dx.$$

Now we simplify the right-hand integrand:

$$\int \sqrt{1+x^2} \, dx = x\sqrt{1+x^2} - \int \frac{(1+x^2)-1}{\sqrt{1+x^2}} \, dx$$

$$= x\sqrt{1+x^2} - \int \frac{1+x^2}{\sqrt{1+x^2}} \, dx + \int \frac{1}{\sqrt{1+x^2}} \, dx$$

$$= x\sqrt{1+x^2} - \int \sqrt{1+x^2} \, dx + \text{arsinh } x + C_1.$$

Now we have the original integrand on the right-hand side, therefore

$$2\int \sqrt{1+x^2} \, dx = x\sqrt{1+x^2} + \text{arsinh } x + C_1$$

$$\int \sqrt{1+x^2} \, dx = \tfrac{1}{2}x\sqrt{1+x^2} + \tfrac{1}{2}\text{arsinh } x + C.$$

Figure 7.21 shows the graphs of $y = \sqrt{1+x^2}$ and $y = \tfrac{1}{2}x\sqrt{1+x^2} + \tfrac{1}{2}\text{arsinh } x$.

Fig. 7.21 The graphs of
$y = \sqrt{1+x^2}$ (broken line)
and $y =$
$\tfrac{1}{2}x\sqrt{1+x^2} + \tfrac{1}{2}\text{arsinh } x$

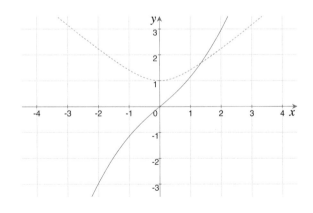

7.4.9 *Integration by Substitution*

Integration by substitution is based upon the chain rule for differentiating a function of a function, which states that if y is a function of u, which in turn is a function of x, then

$$\frac{dy}{dx} = \frac{dy}{du} \cdot \frac{du}{dx}.$$

This integrand

$$\int x^2 \sqrt{x^3}\, dx.$$

is easily solved by rewriting it:

$$\int x^2 \sqrt{x^3}\, dx = \int x^{\frac{7}{2}}\, dx$$
$$= \tfrac{2}{9} x^{\frac{9}{2}} + C.$$

However, introducing a constant term within the square-root requires integration by substitution. For example,

$$\int x^2 \sqrt{x^3 + 1}\, dx.$$

First, we let $u = x^3 + 1$, then

$$\frac{du}{dx} = 3x^2 \quad \text{or} \quad dx = \frac{1}{3x^2} du.$$

Substituting u and dx in the integrand gives

$$\int x^2 \sqrt{x^3 + 1}\, dx = \int x^2 \sqrt{u}\, \frac{1}{3x^2} du$$
$$= \tfrac{1}{3} \int \sqrt{u}\, du$$
$$= \tfrac{1}{3} \int u^{\frac{1}{2}}\, du$$
$$= \tfrac{1}{3} \cdot \tfrac{2}{3} u^{\frac{3}{2}} + C$$
$$= \tfrac{2}{9} (x^3 + 1)^{\frac{3}{2}} + C.$$

Figure 7.22 shows the graphs of $y = x^2 \sqrt{x^3 + 1}$ and $y = \tfrac{2}{9}(x^3 + 1)^{\frac{3}{2}}$.
    Let's try

$$\int 2 \sin x \cdot \cos x\, dx.$$

Fig. 7.22 The graphs of
$y = x^2\sqrt{x^3 + 1}$ (broken
line) and $y = \frac{2}{9}(x^3 + 1)^{\frac{3}{2}}$

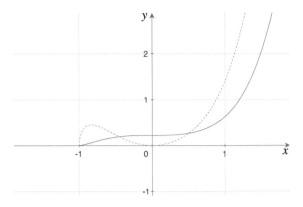

Integrating by substitution we let $u = \sin x$, then

$$\frac{du}{dx} = \cos x \quad \text{or} \quad dx = \frac{1}{\cos x} du.$$

Substituting u and dx in the integrand gives

$$\int 2 \sin x \cdot \cos x \, dx = 2 \int u \cos x \frac{1}{\cos x} du$$

$$= 2 \int u \, du$$

$$= u^2 + C_1$$

$$= \sin^2 x + C.$$

Figure 7.23 shows the graphs of $y = 2 \sin x \cdot \cos x$ and $y = \sin^2 x$.

Fig. 7.23 The graphs of
$y = 2 \sin x \cdot \cos x$ (broken
line) and $y = \sin^2 x$

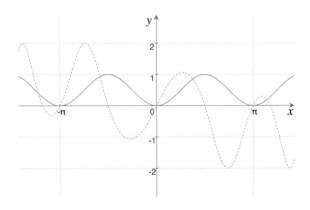

Fig. 7.24 The graphs of
$y = 2e^{\cos(2x)} \sin x \cdot \cos x$
(broken line) and
$y = -\frac{1}{2}e^{\cos(2x)}$

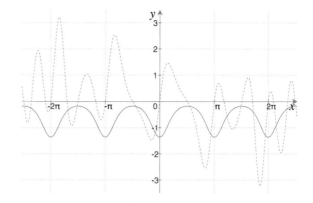

The next example looks difficult, but turns out to be simple:

$$\int 2e^{\cos(2x)} \sin x \cdot \cos x \, dx.$$

Integrating by substitution, let $u = \cos(2x)$, then

$$\frac{du}{dx} = -2\sin(2x) \quad \text{or} \quad dx = -\frac{1}{2\sin(2x)}du.$$

Substituting a double-angle identity, u and du:

$$\int 2e^{\cos(2x)} \sin x \cdot \cos x \, dx = -\int e^u \sin(2x)\frac{1}{2\sin(2x)}du$$

$$= -\frac{1}{2}\int e^u \, du$$

$$= -\frac{1}{2}e^u + C$$

$$= -\frac{1}{2}e^{\cos(2x)} + C.$$

Figure 7.24 shows the graphs of $y = 2e^{\cos(2x)} \sin x \cdot \cos x$ and $y = -\frac{1}{2}e^{\cos(2x)}$.
Now let's try

$$\int \frac{\cos x}{(1 + \sin x)^3} \, dx.$$

Integrating by substitution, let $u = 1 + \sin x$, then

$$\frac{du}{dx} = \cos x \quad \text{or} \quad dx = \frac{1}{\cos x}du.$$

Fig. 7.25 The graphs of
$y = \frac{\cos x}{(1+\sin x)^3}$ (broken line)
and $y = -\frac{1}{2(1+\sin x)^2}$

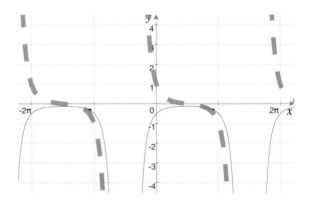

$$\int \frac{\cos x}{(1 + \sin x)^3} \, dx = \int \frac{\cos x}{u^3} \frac{1}{\cos x} du$$

$$= \int u^{-3} \, du$$

$$= -\tfrac{1}{2} u^{-2} + C$$

$$= -\tfrac{1}{2} (1 + \sin x)^{-2} + C$$

$$= -\frac{1}{2(1 + \sin x)^2} + C.$$

Figure 7.25 shows the graphs of $y = \frac{\cos x}{(1+\sin x)^3}$ and $y = -\frac{1}{2(1+\sin x)^2}$.
Finally, an easy one:

$$\int \sin(2x) \, dx.$$

Integrating by substitution, let $u = 2x$, then

$$\frac{du}{dx} = 2 \quad \text{or} \quad dx = \tfrac{1}{2} du.$$

$$\int \sin(2x) \, dx = \tfrac{1}{2} \int \sin u \, du$$

$$= -\tfrac{1}{2} \cos u + C$$

$$= -\tfrac{1}{2} \cos(2x) + C$$

Figure 7.26 shows the graphs of $y = \sin(2x)$ and $y = -\tfrac{1}{2} \cos(2x)$.

Fig. 7.26 The graphs of
$y = \sin(2x)$ (broken line)
and $y = -\frac{1}{2}\cos(2x)$

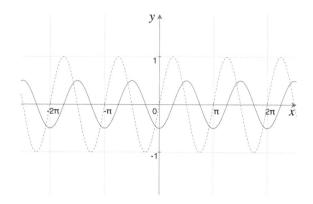

7.4.10 Partial Fractions

Integration by *partial fractions* is used when an integrand's denominator contains
a product that can be split into two fractions. For example, it should be possible to
convert

$$\int \frac{3x + 4}{(x + 1)(x + 2)}\, dx$$

into

$$\int \frac{A}{x + 1}\, dx + \int \frac{B}{x + 2}\, dx$$

which individually, are easy to integrate. Let's compute A and B:

$$\frac{3x + 4}{(x + 1)(x + 2)} = \frac{A}{x + 1} + \frac{B}{x + 2}$$
$$3x + 4 = A(x + 2) + B(x + 1)$$
$$= Ax + 2A + Bx + B.$$

Equating constants and terms in x:

$$4 = 2A + B \qquad\qquad (7.1)$$
$$3 = A + B \qquad\qquad (7.2)$$

Subtracting (7.2) from (7.1), gives $A = 1$ and $B = 2$. Therefore,

$$\int \frac{3x + 4}{(x + 1)(x + 2)}\, dx = \int \frac{1}{x + 1}\, dx + \int \frac{2}{x + 2}\, dx$$
$$= \ln(x + 1) + 2\ln(x + 2) + C.$$

Fig. 7.27 The graphs of
$y = \frac{3x+4}{(x+1)(x+2)}$ (broken line)
and
$y = \ln(x + 1) + 2\ln(x + 2)$

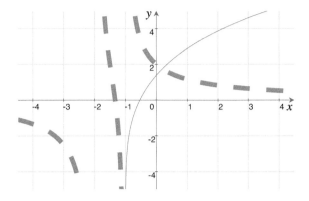

Figure 7.27 shows the graphs of $y = \frac{3x+4}{(x+1)(x+2)}$ and $y = \ln(x + 1) + 2\ln(x + 2)$.
Now let's try

$$\int \frac{5x - 7}{(x - 1)(x - 2)}\, dx.$$

Integrating by partial fractions:

$$\frac{5x - 7}{(x - 1)(x - 2)} = \frac{A}{x - 1} + \frac{B}{x - 2}$$
$$5x - 7 = A(x - 2) + B(x - 1)$$
$$= Ax + Bx - 2A - B.$$

Equating constants and terms in x:

$$-7 = -2A - B \tag{7.3}$$
$$5 = A + B \tag{7.4}$$

Subtracting (7.3) from (7.4), gives $A = 2$ and $B = 3$. Therefore,

$$\int \frac{3x + 4}{(x - 1)(x - 2)}\, dx = \int \frac{2}{x - 1}\, dx + \int \frac{3}{x - 2}\, dx$$
$$= 2\ln(x - 1) + 3\ln(x - 2) + C.$$

Figure 7.28 shows the graphs of $y = \frac{5x-7}{(x-1)(x-2)}$ and $y = 2\ln(x - 1) + 3\ln(x - 2)$.

Fig. 7.28 The graphs of
$y = \frac{5x-7}{(x-1)(x-2)}$ (broken line)
and $y =$
$2\ln(x-1) + 3\ln(x-2)$

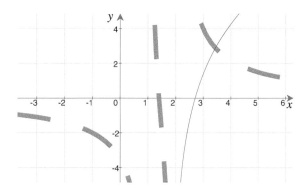

The next example requires fractions:

$$\int \frac{6x^2 + 5x - 2}{x^3 + x^2 - 2x}\, dx.$$

Integrating by partial fractions:

$$\frac{6x^2 + 5x - 2}{x^3 + x^2 - 2x} = \frac{A}{x} + \frac{B}{x+2} + \frac{C}{x-1}$$
$$6x^2 + 5x - 2 = A(x+2)(x-1) + Bx(x-1) + Cx(x+2)$$
$$= Ax^2 + Ax - 2A + Bx^2 - Bx + Cx^2 + 2Cx.$$

Equating constants, terms in x and x^2:

$$-2 = -2A \tag{7.5}$$
$$5 = A - B + 2C \tag{7.6}$$
$$6 = A + B + C \tag{7.7}$$

Manipulating (7.5), (7.6) and (7.7): $A = 1$, $B = 2$ and $C = 3$, therefore,

$$\int \frac{6x^2 + 5x - 2}{x^3 + x^2 - 2x}\, dx = \int \frac{1}{x}\, dx + \int \frac{2}{x+2}\, dx + \int \frac{3}{x-1}\, dx$$
$$= \ln x + 2\ln(x+2) + 3\ln(x-1) + C.$$

Figure 7.29 shows the graphs of $y = \frac{6x^2+5x-2}{x^3+x^2-2x}$ and $y = \ln x + 2\ln(x+2) + 3\ln(x-1)$.

Fig. 7.29 The graphs of
$y = \frac{6x^2+5x-2}{x^3+x^2-2x}$ (broken line)
and $y = \ln x + 2\ln(x + 2) + 3\ln(x - 1)$

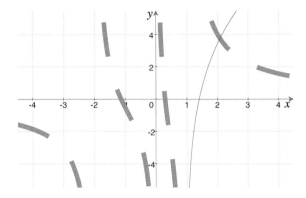

7.5 Summary

This chapter introduced a collection of strategies that should be considered when integrating a function. It is far from complete, and one must expect that some integrands will prove extremely difficult to solve, and software has to be used to reveal a numerical solution.

Chapter 8
Area Under a Graph

8.1 Introduction

The ability to calculate the area under a graph is one of the most important discoveries of integral Calculus. Prior to Calculus, area was computed by dividing a zone into very small strips and summing the individual areas. The accuracy of the result is improved simply by making the strips smaller and smaller, taking the result towards some limiting value. In this chapter I show how integral Calculus provides a way to compute the area between a function's graph and the x- and y-axis.

8.2 Calculating Areas

Before considering the relationship between area and integration, let's see how area is calculated using functions and simple geometry.

Figure 8.1 shows the graph of $y = 1$, where the area $A(x)$ of the shaded zone is

$$A(x) = x, \quad x > 0.$$

For example, $A(4) = 4$, and $A(10) = 10$. An interesting observation is that the derivative of $A(x)$ is the equation of the line:

$$\frac{dA}{dx} = 1 = y.$$

Figure 8.2 shows the graph of $y = 2x$. The area $A(x)$ of the shaded triangle is

$$A(x) = \tfrac{1}{2} base \cdot height$$
$$= \tfrac{1}{2}x \cdot 2x$$
$$= x^2.$$

© Springer Nature Switzerland AG 2019
J. Vince, *Calculus for Computer Graphics*,
https://doi.org/10.1007/978-3-030-11376-6_8

Fig. 8.1 Area of the shaded
zone is $A(x) = x$

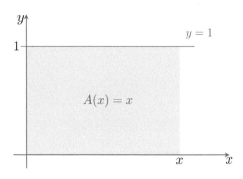

Fig. 8.2 Area of the shaded
zone is $A(x) = x^2$

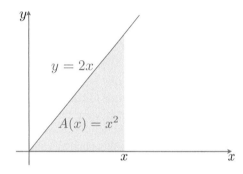

Fig. 8.3 Graph of
$y = \sqrt{r^2 - x^2}$

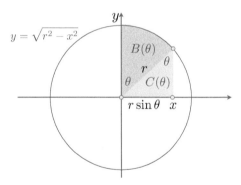

Thus, $A(4) = 16$, and $A(5) = 25$. Once again, the derivative of $A(x)$ is the equation
of the line:
$$\frac{dA}{dx} = 2x = y$$

which is no coincidence.

Finally, Fig. 8.3 shows a circle where $x^2 + y^2 = r^2$, and the curve of the first
quadrant is described by the function

$$y = \sqrt{r^2 - x^2}, \quad x \in [0, r]$$

The total area of the shaded zones is the sum of the two parts: $B(\theta)$ and $C(\theta)$. The function is defined in terms of the angle θ, such that

$$x = r \sin \theta$$
$$y = r \cos \theta.$$

Therefore,

$$B(\theta) = \tfrac{1}{2} r^2 \theta$$
$$C(\theta) = \tfrac{1}{2}(r \cos \theta)(r \sin \theta) = \tfrac{1}{4} r^2 \sin(2\theta)$$
$$A(\theta) = B(\theta) + C(\theta)$$
$$= \tfrac{1}{2} r^2 \left(\theta + \tfrac{1}{2} \sin(2\theta)\right).$$

Differentiating $A(\theta)$:

$$\frac{dA}{d\theta} = \tfrac{1}{2} r^2 \left(1 + \cos(2\theta)\right) = r^2 \cos^2 \theta.$$

But we want the derivative with respect to x, which requires the chain rule:

$$\frac{dA}{dx} = \frac{dA}{d\theta} \cdot \frac{d\theta}{dx}$$

where

$$\frac{dx}{d\theta} = r \cos \theta$$

or

$$\frac{d\theta}{dx} = \frac{1}{r \cos \theta}$$

therefore

$$\frac{dA}{dx} = \frac{r^2 \cos^2 \theta}{r \cos \theta} = r \cos \theta = y$$

which is the equation for the quadrant. When $\theta = \pi/2$, $A(\pi/2)$ equals the area of a quadrant of a unit-radius circle:

$$A\left(\tfrac{\pi}{2}\right) = \tfrac{1}{2} r^2 \left(\theta + \tfrac{1}{2} \sin(2\theta)\right)$$
$$= \tfrac{1}{2} \left(\tfrac{1}{2}\pi + \tfrac{1}{2} \sin \pi\right)$$
$$= \tfrac{1}{2} \left(\tfrac{1}{2}\pi\right)$$
$$= \tfrac{1}{4}\pi$$

and the area of a unit-radius circle is four quadrants: $A(\theta) = \pi$.

Fig. 8.4 Relationship
between $y = f(x)$ and $A(x)$

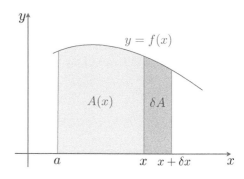

Fig. 8.4 Relationship
between $y = f(x)$ and $A(x)$

Hopefully, these three examples provide strong evidence that the derivative of the function for the area under a graph, equals the graph's function:

$$\frac{dA}{dx} = f(x)$$

and

$$dA = f(x)\, dx$$

which implies that

$$A = \int f(x)\, dx.$$

Now let's prove this observation using Fig. 8.4, which shows a continuous function $y = f(x)$. Next, we define a function $A(x)$ to represent the area under the graph over the interval $[a, x]$. δA is the area increment between x and $x + \delta x$, and

$$\delta A \approx f(x) \cdot \delta x.$$

We can also reason that

$$\delta A = A(x + \delta x) - A(x) \approx f(x) \cdot \delta x$$

and the derivative dA/dx is the limiting condition

$$\frac{dA}{dx} = \lim_{\delta x \to 0} \frac{A(x + \delta x) - A(x)}{\delta x} = \lim_{\delta x \to 0} \frac{f(x) \cdot \delta x}{\delta x} = f(x)$$

thus,

$$\frac{dA}{dx} = f(x).$$

Fig. 8.5 $A(b)$ is the area under the graph $y = f(x)$, $x \in [0, b]$

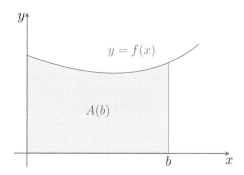

This can be rearranged as

$$dA = f(x)\, dx$$

whose antiderivative is

$$A(x) = \int f(x)\, dx.$$

The function $A(x)$ computes the area over the interval $x \in [a, b]$ and is represented by

$$A(x) = \int_a^b f(x)\, dx$$

which is called *the integral* or *definite integral*.

Let's assume that $A(b)$ is the area under the graph of $f(x)$ over the interval $x \in [0, b]$, as shown in Fig. 8.5, and is written

$$A(b) = \int_0^b f(x)\, dx.$$

Similarly, let $A(a)$ be the area under the graph of $f(x)$ over the interval $x \in [0, a]$, as shown in Fig. 8.6, and is written

$$A(a) = \int_0^a f(x)\, dx.$$

Figure 8.7 shows that the area of the shaded zone over the interval $x \in [a, b]$ is calculated by

$$A(b) - A(a)$$

which is written

$$A(b) - A(a) = \int_0^b f(x)\, dx - \int_0^a f(x)\, dx$$

Fig. 8.6 $A(a)$ is the area under the graph $y = f(x)$, $x \in [0, a]$

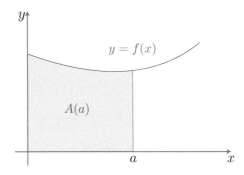

Fig. 8.7 $A(b) - A(a)$ is the area under the graph $y = f(x), x \in [a, b]$

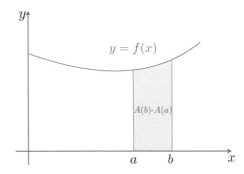

and is contracted to

$$A(b) - A(a) = \int_a^b f(x)\, dx. \qquad (8.1)$$

The *fundamental theorem of Calculus* states that the definite integral

$$\int_a^b f(x)\, dx = F(b) - F(a)$$

where

$$F(a) = \int f(x)\, dx, \quad x = a$$

$$F(b) = \int f(x)\, dx, \quad x = b.$$

In order to compute the area beneath a graph of $f(x)$ over the interval $x \in [a, b]$, we first integrate the graph's function

$$F(x) = \int f(x)\, dx$$

Fig. 8.8 Area under the
graph is $\int_1^4 1\, dx$

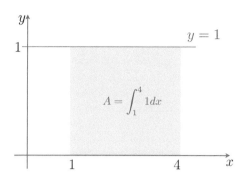

and then calculate the area, which is the difference

$$F(b) - F(a).$$

Let's show how (8.1) is used in the context of the earlier three examples.

We start by calculating the area under $y = 1$, over the interval $x \in [1, 4]$, as shown
in Fig. 8.8. Beginning with

$$A = \int_1^4 1\, dx.$$

Next, we integrate the function, and transfer the interval bounds employing the *substitution symbol* $\Big|_1^4$, or square brackets $\Big[\ \Big]_1^4$. Using $\Big|_1^4$, we have

$$A = x\ \Big|_1^4$$
$$= 4 - 1$$
$$= 3$$

or using $\Big[\ \Big]_1^4$, we have

$$A = \Big[x\Big]_1^4$$
$$= 4 - 1$$
$$= 3.$$

Next, we calculate the area under $y = 2x$, over the interval $x \in [1, 4]$, as shown
in Fig. 8.9. We begin with

$$A = \int_1^4 2x\, dx$$

Fig. 8.9 Area under the graph is $\int_1^4 2x \, dx$

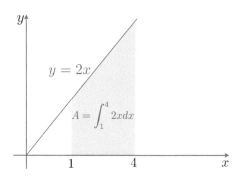

and integrate the function and evaluate the area

$$A = x^2 \Big|_1^4$$
$$= 16 - 1$$
$$= 15.$$

Last, we calculate the area under $y = \sqrt{r^2 - x^2}$, over the interval $x \in [0, r]$, which is the equation for the quadrant of a circle, as shown in Fig. 8.3. We begin with

$$A = \int_0^r \sqrt{r^2 - x^2} \, dx. \tag{8.2}$$

Unfortunately, (8.2) contains a function of a function, which is resolved by substituting another independent variable. In this case, the geometry of the circle suggests

$$x = r \sin \theta$$

therefore,

$$\sqrt{r^2 - x^2} = r \cos \theta$$

and

$$\frac{dx}{d\theta} = r \cos \theta. \tag{8.3}$$

However, changing the independent variable requires changing the interval for the integral. In this case, changing $x \in [0, r]$ into $\theta \in [\theta_1, \theta_2]$:

When $x = 0$, $r \sin \theta_1 = 0$, therefore $\theta_1 = 0$.
When $x = r$, $r \sin \theta_2 = r$, therefore $\theta_2 = \pi/2$.
Thus, the new interval is $\theta \in [0, \pi/2]$.
Finally, the dx in (8.2) has to be changed into $d\theta$, which using (8.3) makes

Fig. 8.10 The two areas
associated with a sine wave

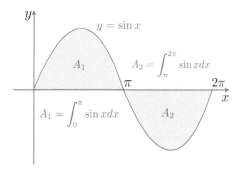

$$dx = r \cos \theta \, d\theta.$$

We are now in a position to rewrite the original integral using θ as the independent variable:

$$A = \int_0^{\frac{\pi}{2}} (r \cos \theta)(r \cos \theta) \, d\theta$$

$$= r^2 \int_0^{\frac{\pi}{2}} \cos^2 \theta \, d\theta$$

$$= \tfrac{1}{2} r^2 \int_0^{\frac{\pi}{2}} 1 + \cos(2\theta) \, d\theta$$

$$= \tfrac{1}{2} r^2 \left[\theta + \tfrac{1}{2} \sin(2\theta) \right]_0^{\frac{\pi}{2}}$$

$$= \tfrac{1}{2} r^2 \cdot \tfrac{1}{2} \pi$$

$$= \tfrac{1}{4} \pi r^2$$

which makes the area of a full circle πr^2.

8.3 Positive and Negative Areas

Area in the real world is always a positive quantity—no matter how it is measured. In Calculus, however, the integral is a signed quantity, such that areas above the x-axis are positive, whilst areas below the x-axis are negative. This can be illustrated by computing the area of the positive and negative parts of a sine wave.

Figure 8.10 shows a sine wave over one cycle, where the area above the x-axis is labelled A_1, and the area below the x-axis is labelled A_2. These areas are computed as follows.

Fig. 8.11 The accumulated
area of a sine wave

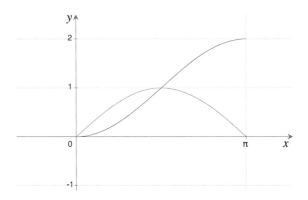

$$A_1 = \int_0^\pi \sin x \, dx$$
$$= -\cos x \Big|_0^\pi$$
$$= 1 + 1$$
$$= 2.$$

However, A_2 gives a negative result:

$$A_2 = \int_\pi^{2\pi} \sin x \, dx$$
$$= -\cos x \Big|_\pi^{2\pi}$$
$$= -1 - 1$$
$$= -2.$$

This means that the area is zero over the interval $x \in [0, 2\pi]$. Consequently, one must be very careful using this technique for functions that are negative in the interval under investigation.

Figure 8.11 shows $\sin x$ over the interval $x \in [0, \pi]$ and its accumulated area.

8.4   Area Between Two Functions

Figure 8.12 shows the graphs of $y = x^2$ and $y = x^3$, with two areas labelled A_1 and A_2. A_1 is the area trapped between the two graphs over the interval $x \in [-1, 0]$ and A_2 is the area trapped between the two graphs over the interval $x \in [0, 1]$. These areas are calculated very easily: in the case of A_1 we sum the individual areas under

Fig. 8.12 Two areas between $y = x^2$ and $y = x^3$

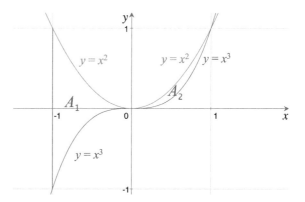

the two graphs, remembering to reverse the sign for the area associated with $y = x^3$. For A_2 we subtract the individual areas under the two graphs.

$$A_1 = \int_{-1}^{0} x^2 \, dx - \int_{-1}^{0} x^3 \, dx$$
$$= \tfrac{1}{3} x^3 \Big|_{-1}^{0} - \tfrac{1}{4} x^4 \Big|_{-1}^{0}$$
$$= \tfrac{1}{3} + \tfrac{1}{4}$$
$$= \tfrac{7}{12}.$$

$$A_2 = \int_{0}^{1} x^2 \, dx - \int_{0}^{1} x^3 \, dx$$
$$= \tfrac{1}{3} x^3 \Big|_{0}^{1} - \tfrac{1}{4} x^4 \Big|_{0}^{1}$$
$$= \tfrac{1}{3} - \tfrac{1}{4}$$
$$= \tfrac{1}{12}.$$

Note, that in both cases the calculation is the same, which implies that when we employ

$$A = \int_{a}^{b} \Big(f(x) - g(x) \Big) \, dx$$

A is always the area trapped between $f(x)$ and $g(x)$ over the interval $x \in [a, b]$.

Let's take another example, by computing the area A between $y = \sin x$ and the line $y = \tfrac{1}{2}$, as shown in Fig. 8.13. The horizontal line intersects the sine curve at $x = 30°$ and $x = 150°$, marked in radians as 0.5236 and 2.618 respectively.

Fig. 8.13 The area between
$y = \sin x$ and $y = 0.5$

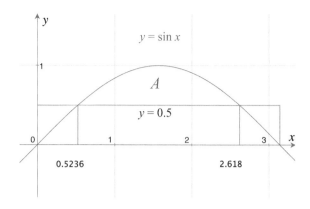

$$A = \int_{\pi/6}^{5\pi/6} \sin x \, dx - \int_{\pi/6}^{5\pi/6} \tfrac{1}{2} \, dx$$

$$= -\cos x \, \Big|_{\pi/6}^{5\pi/6} - \tfrac{1}{2}x \, \Big|_{\pi/6}^{5\pi/6}$$

$$= \left(\frac{\sqrt{3}}{2} + \frac{\sqrt{3}}{2} \right) - \tfrac{1}{2} \left(\frac{5\pi}{6} - \frac{\pi}{6} \right)$$

$$= \sqrt{3} - \tfrac{1}{3}\pi$$

$$\approx 0.685.$$

8.5   Areas with the y-Axis

So far we have only calculated areas between a function and the x-axis. So let's
compute the area between a function and the y-axis. Figure 8.14 shows the function
$y = x^2$ over the interval $x \in [0, 4]$, where A_1 is the area between the curve and the
x-axis, and A_2 is the area between the curve and y-axis. The sum $A_1 + A_2$ must
equal $4 \times 16 = 64$, which is a useful control. Let's compute A_1.

$$A_1 = \int_0^4 x^2 \, dx$$

$$= \tfrac{1}{3}x^3 \, \Big|_0^4$$

$$= \frac{64}{3}$$

$$\approx 21.333$$

Fig. 8.14 The areas between
the x-axis and the y-axis

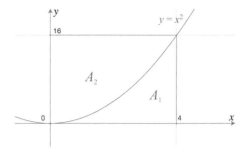

which means that $A_2 \approx 42.666$. To compute A_2 we construct an integral relative to dy with a corresponding interval. If $y = x^2$ then $x = y^{\frac{1}{2}}$, and the interval is $y \in [0, 16]$:

$$A_2 = \int_0^{16} y^{\frac{1}{2}} \, dy$$

$$= \tfrac{2}{3} y^{\frac{3}{2}} \Big|_0^{16}$$

$$= \tfrac{2}{3} 64$$

$$\approx 42.666.$$

8.6   Area with Parametric Functions

When working with functions of the form $y = f(x)$, the area under its curve and the x-axis over the interval $x \in [a, b]$ is

$$A = \int_a^b f(x) \, dx.$$

However, if the curve has a parametric form where

$$x = f_x(t) \quad \text{and} \quad y = f_y(t)$$

then we can derive an equivalent integral as follows.

First, we need to establish equivalent limits $[\alpha, \beta]$ for t, such that

$$a = f_x(\alpha) \quad \text{and} \quad b = f_y(\beta).$$

Second, any point on the curve has corresponding Cartesian and parametric coordinates:

$$x \quad \text{and} \quad f_x(t)$$

$$y = f(x) \quad \text{and} \quad f_y(t).$$

Third,

$$x = f_x(t)$$
$$dx = f_x'(t)\, dt$$
$$A = \int_a^b f(x)\, dx$$
$$= \int_\alpha^\beta f_y(t) \cdot f_x'(t)\, dt$$

therefore

$$A = \int_\alpha^\beta f_y(t) \cdot f_x'(t)\, dt. \tag{8.4}$$

Let's apply (8.4) using the parametric equations for a circle

$$x = -r \cos t$$
$$y = r \sin t.$$

as shown in Fig. 8.15. Remember that the Cartesian interval is $[a, b]$ left to right, and the polar interval $[\alpha, \beta]$, must also be left to right, which is why $x = -r \cos t$. Therefore,

$$f_x'(t) = r \sin t$$
$$f_y(t) = r \sin t$$
$$A = \int_\alpha^\beta f_y(t) \cdot f_x'(t)\, dt$$
$$= \int_0^\pi r \sin t \cdot r \sin t\, dt$$
$$= r^2 \int_0^\pi \sin^2 t\, dt$$
$$= \tfrac{1}{2} r^2 \int_0^\pi 1 - \cos(2t)\, dt$$
$$= \tfrac{1}{2} r^2 \left[t + \tfrac{1}{2} \sin(2t) \right]_0^\pi$$
$$= \tfrac{1}{2} \pi r^2$$

which makes the area of a full circle πr^2.

Fig. 8.15 The parametric functions for a circle

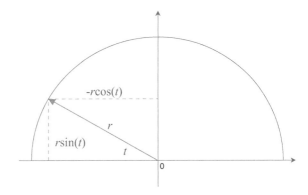

8.7 Bernhard Riemann

The German mathematician Bernhard Riemann (1826–1866) (pronounced 'Ree-man') made major contributions to various areas of mathematics, including integral Calculus, where his name is associated with a formal method for summing areas and volumes. Through the *Riemann Sum*, Riemann provides an elegant and consistent notation for describing single, double and triple integrals when calculating area and volume. I will show how the Riemann Sum explains why the area under a curve is the function's integral. But first, I need to explain some incidental notation used in the description.

8.7.1 Domains and Intervals

Consider any continuous, real-valued function $f(x)$ which returns a meaningful value for a wide range of x-values. For example, the function $f(x) = x^2$ works with any negative or positive x. This is called the *domain* of $f(x)$ and written using *interval* notation as $(-\infty, \infty)$, where the parentheses () remind us not to include $-\infty$ and ∞ in the domain, as they have no definite value. When we wish to focus upon a specific domain such as $a \leq x \leq b$, then we write $[a, b]$, where the square brackets remind us to include a and b in the domain. The function $f(x) = \sqrt{x}$ returns a real value, so long as $x \geq 0$, which means that its domain is $[0, \infty)$.

Some functions, like $f(x) = 1/(x - 2)$ are sensitive to just one value—in this case when $x = 2$—which creates a divide by zero. Therefore, there are two intervals: $(-\infty, 2)$ and $(2, \infty)$, which in set notation is written

$$(-\infty, 2) \cup (2, \infty).$$

We are normally at liberty to choose the domain of a function—provided that we can actually compute it. The domain then becomes part of the definition of a function.

8.7.2   The Riemann Sum

Figure 8.16 shows a function $f(x)$ divided into eight equal sub-intervals where

$$\Delta x = \frac{b-a}{8}$$

and

$$a = x_0 < x_1 < x_2 < \cdots < x_7 < x_8 = b.$$

In order to compute the area under the curve over the interval $[a, b]$, the interval is divided into some large number of sub-intervals. In this case, eight, which is not very large, but convenient to illustrate. Each sub-interval becomes a rectangle with a common width Δx and a different height. The area of the first rectangular sub-interval shown shaded, can be calculated in various ways. We can take the left-most height x_0 and form the product $x_0 \Delta x$, or we can take the right-most height x_1 and form the product $x_1 \Delta x$. On the other hand, we could take the mean of the two heights $(x_0 + x_1)/2$ and form the product $(x_0 + x_1)\Delta x/2$. A solution that shows no bias towards either left, right or centre, is to let x_i^* be anywhere in a specific sub-interval Δx_i, then the area of the rectangle associated with the sub-interval is $f(x_i^*)\Delta x_i$, and the sum of the rectangular areas is given by

$$A = \sum_{i=1}^{8} f(x_i^*)\Delta x_i.$$

Dividing the interval into eight equal sub-intervals will not generate a very accurate result for the area under the graph. But increasing it to eight-thousand or eight-million, will take us towards some limiting value. Rather than specify some specific large number, it is common practice to employ n, and let n tend towards infinity, which is written

$$A = \sum_{i=1}^{n} f(x_i^*)\Delta x_i. \tag{8.5}$$

The right-hand side of (8.5) is called a Riemann Sum, of which there are many. For the above description, I have assumed that the sub-intervals are equal, which is not a necessary requirement.

If the number of sub-intervals is n, then

$$\Delta x = \frac{b-a}{n}$$

and the definite integral is defined as

Fig. 8.16 The graph of function $f(x)$ over the interval $[a, b]$

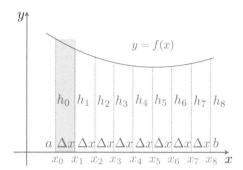

$$\int_a^b f(x)\, dx = \lim_{n \to \infty} \sum_{i=1}^{n} f(x_i^*)\Delta x_i.$$

In later chapters, double and triple integrals are used to compute areas and volumes, and require us to think carefully about their meaning and what they are doing. Dividing space into sub-intervals, sub-areas or sub-volumes, provides a consistent strategy for increasing our understanding of the subject.

8.8 Summary

In this chapter we have discovered the double role of integration. Integrating a function reveals another function, whose derivative is the function under investigation. Simultaneously, integrating a function computes the area between the function's graph and the x- or y-axis. Although the concept of area in every-day life is an unsigned quantity, within mathematics, and in particular Calculus, area *is* a signed quality, and one must be careful when making such calculations.

Chapter 9
Arc Length and Parameterisation of Curves

9.1 Introduction

In previous chapters we have seen how Calculus reveals the slope and the area under a function's graph, and it should be no surprise that it can be used to compute the arc length of a continuous function. However, although the formula for the arc length results in a simple integrand, it is not always possible to integrate, and other numerical techniques have to be used.

Vector-valued functions are widely used for curve generation, and we explore strategies for computing the arc lengths of a circle, parabola, ellipse, hyperbolic cosh, helix, 2D and 3D quadratic Bézier curves. We then investigate the arc-length parameterisation of a 3D line and helix curve, and show how points are positioned on these using a square law distribution. Finally, I show how to deal with functions expressed in polar coordinates. In order to compute a function's arc length using integration, we first need to understand the *mean-value theorem*.

9.2 Lagrange's Mean-Value Theorem

The French mathematician Joseph Louis Lagrange (1736–1813) is acknowledged as being the first person to state the mean-value theorem:

A function $f(x)$ that is continuous in the closed interval $[a, b]$ and differentiable in the open interval $]a, b[$ has in this interval at least one value c such that $f'(c)$ equals

$$f'(c) = \frac{f(b) - f(a)}{b - a}.$$

Figure 9.1 illustrates the geometry behind this theorem, where we see the graph of a function $f(x)$, which has no discontinuities over the interval $x \in [a, b]$. Although

© Springer Nature Switzerland AG 2019
J. Vince, *Calculus for Computer Graphics*,
https://doi.org/10.1007/978-3-030-11376-6_9

Fig. 9.1 The secant's slope
equals the tangent

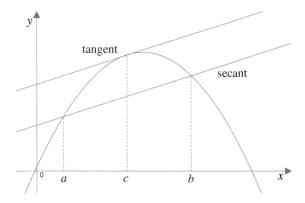

not shown, we assume that the function is differentiable outside the bounds of
the interval. The slope of the line (secant) connecting the points $(a, f(a))$ and
$(b, f(b))$ is

$$\frac{f(b) - f(a)}{b - a}$$

and the mean-value theorem states that this slope equals the tangent at another point
c, where $a < c < b$. One can easily visualise this from Fig. 9.1 by tracking the slope
of $f(x)$ over the interval $x \in [a, b]$. At $x = a$, the slope, given by $f'(a)$, has some
positive value, whereas at $x = b$, the slope, given by $f'(b)$, has some negative value.
Clearly, the secant's slope is less than $f'(a)$ and greater than $f'(b)$ and must equal
$f'(c)$, somewhere between a and b. Lagrange provided a rigorous mathematical proof
for any function within the constraints of the theorem. We call upon this theorem in
the next section.

9.3   Arc Length

In every-day life we can measure the length of a curved surface by laying a flexi-
ble tape measure upon it and taking a reading. Given the graph of a mathematical
function, we can measure its length by reducing it to a chain of straight lines and
summing their individual lengths. Although this is rather crude, accuracy is improved
by making the straight lines increasingly shorter. This is the approach we employ in
the following analysis.

Figure 9.2 shows part of a curve divided into n intervals where any sample point
P_i has coordinates (x_i, y_i), where $0 < i < n$. Using the theorem of Pythagoras, the
distance between two points P_i and P_{i+1} is given by

$$\Delta s = \sqrt{(x_{i+1} - x_i)^2 + (y_{i+1} - y_i)^2}$$
$$= \sqrt{(\Delta x_i)^2 + (\Delta y_i)^2}$$

Fig. 9.2 The chain of
straight-line segments
approximates to the
curve's length

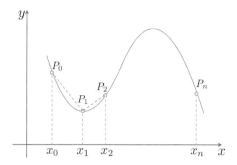

and the approximate length between P_0 and P_n is given by

$$s \approx \sum_{i=1}^{n} \sqrt{(\Delta x_i)^2 + (\Delta y_i)^2}.$$

As n tends towards infinity, then in the limit

$$s = \lim_{n \to \infty} \sum_{i=1}^{n} \sqrt{(\Delta x_i)^2 + (\Delta y_i)^2}$$

$$= \lim_{n \to \infty} \sum_{i=1}^{n} \sqrt{1 + \left(\frac{\Delta y_i}{\Delta x_i}\right)^2} \, \Delta x_i \tag{9.1}$$

Lagrange's mean-value theorem states that there must be a value x_j, such that $x_{i-1} < x_j < x_i$, where

$$f'(x_j) = \frac{f(x_i) - f(x_{i-1})}{x_i - x_{i-1}}$$

$$= \frac{y_i - y_{i-1}}{x_i - x_{i-1}}$$

$$= \frac{\Delta y_i}{\Delta x_i}.$$

Therefore, (9.1) becomes

$$s = \lim_{n \to \infty} \sum_{i=1}^{n} \sqrt{1 + \left(f'(x_j)\right)^2} \, \Delta x_i$$

and

$$s = \int_{a}^{b} \sqrt{1 + \left(\frac{dy}{dx}\right)^2} \, dx, \quad x \in [a, b]. \tag{9.2}$$

9.3.1 Arc Length of a Straight Line

Let's test (9.2) by finding the length of the straight line $y = \frac{3}{4}x$, over the interval $x \in [0, 4]$, which using simple geometry is 5.

$$\frac{dy}{dx} = \frac{3}{4}$$

therefore,

$$
\begin{aligned}
s &= \int_0^4 \sqrt{1 + \left(\frac{dy}{dx}\right)^2}\, dx \\
&= \int_0^4 \sqrt{1 + \left(\frac{3}{4}\right)^2}\, dx \\
&= \int_0^4 \sqrt{\frac{25}{16}}\, dx \\
&= \int_0^4 \frac{5}{4}\, dx \\
&= \frac{5}{4}x \Big|_0^4 \\
&= 5.
\end{aligned}
$$

9.3.2 Arc Length of a Circle

Figure 9.3 shows a semi-circle with radius r, where $y = \sqrt{r^2 - x^2}$. Therefore,

Fig. 9.3 A semi-circle with radius r

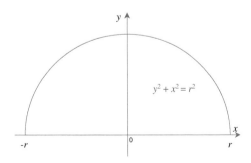

$$\frac{dy}{dx} = \tfrac{1}{2}(r^2 - x^2)^{-1/2} \cdot (-2x)$$

$$= \frac{-x}{\sqrt{r^2 - x^2}}$$

$$\left(\frac{dy}{dx}\right)^2 = \frac{x^2}{r^2 - x^2}.$$

Integrating over the interval $x \in [-r, r]$, which is doubled to give the circle's circumference:

$$s = 2 \int_{-r}^{r} \sqrt{1 + \left(\frac{dy}{dx}\right)^2} \, dx$$

$$= 2 \int_{-r}^{r} \sqrt{1 + \frac{x^2}{r^2 - x^2}} \, dx$$

$$= 2 \int_{-r}^{r} \sqrt{\frac{r^2}{r^2 - x^2}} \, dx$$

$$= 2r \int_{-r}^{r} \frac{dx}{\sqrt{r^2 - x^2}}$$

$$= 2r \arcsin\left(\frac{x}{r}\right)\Big|_{-r}^{r}$$

$$= 2r \left(\tfrac{1}{2}\pi + \tfrac{1}{2}\pi\right)$$

$$= 2\pi r.$$

9.3.3 Arc Length of a Parabola

Let's compute the arc length of the parabola $y = 0.5x^2$, over the interval $x \in [0, 4]$, where $dy/dx = x$:

$$s = \int_{0}^{4} \sqrt{1 + \left(\frac{dy}{dx}\right)^2} \, dx$$

$$= \int_{0}^{4} \sqrt{1 + x^2} \, dx.$$

To remove the radical we let $x = \tan\theta$ where $dx/d\theta = \sec^2\theta$ and continue with an indefinite integral. Therefore,

$$s = \int \sqrt{1 + \tan^2 \theta} \cdot \sec^2 \theta \; d\theta$$

$$= \int \sqrt{\sec^2 \theta} \cdot \sec^2 \theta \; d\theta$$

$$= \int \sec \theta \cdot \sec^2 \theta \; d\theta.$$

Having removed the radical, we are now left with a product, which is integrated by parts, by letting

$$u = \sec \theta \quad \text{and} \quad v' = \sec^2 \theta,$$

which means that

$$u' = \sec \theta \cdot \tan \theta \quad \text{and} \quad v = \tan \theta.$$

Therefore,

$$\int u v' \; d\theta = uv - \int v u' \; d\theta$$

$$\int \sec \theta \cdot \sec^2 \theta \; d\theta = \sec \theta \cdot \tan \theta - \int \tan \theta \cdot \sec \theta \cdot \tan \theta \; d\theta$$

$$= \sec \theta \cdot \tan \theta - \int \sec \theta \cdot \tan^2 \theta \; d\theta$$

$$= \sec \theta \cdot \tan \theta - \int \sec \theta \cdot (\sec^2 \theta - 1) \; d\theta$$

$$= \sec \theta \cdot \tan \theta - \int \sec^3 \theta \; d\theta + \int \sec \theta \; d\theta$$

$$2 \int \sec^3 \theta \; d\theta = \sec \theta \cdot \tan \theta + \int \sec \theta \; d\theta$$

$$\int \sec^3 \theta \; d\theta = \tfrac{1}{2} \sec \theta \cdot \tan \theta + \tfrac{1}{2} \int \sec \theta \; d\theta$$

$$= \tfrac{1}{2} \sec \theta \cdot \tan \theta + \tfrac{1}{2} \ln \left| \sec \theta + \tan \theta \right| + C.$$

Now let's convert this result back to the original function where $x = \tan \theta$ and $\sec \theta = \sqrt{1 + x^2}$ and reintroduce the limits $[0, 4]$:

$$\tfrac{1}{2} \sec \theta \cdot \tan \theta + \tfrac{1}{2} \ln \left| \sec \theta + \tan \theta \right| + C = \tfrac{1}{2} x \sqrt{1 + x^2} + \tfrac{1}{2} \ln \left| \sqrt{1 + x^2} + x \right| + C$$

therefore

$$\int_0^4 \sqrt{1 + x^2} \; dx = \tfrac{1}{2} x \sqrt{1 + x^2} + \tfrac{1}{2} \ln \left| \sqrt{1 + x^2} + x \right| \; \Big|_0^4.$$

Fig. 9.4 Graph of $y = 0.5x^2$

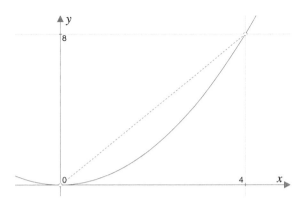

Evaluating this result, we get

$$\int_0^4 \sqrt{1+x^2}\, dx = \tfrac{1}{2}x\sqrt{1+x^2} + \tfrac{1}{2}\ln\left|\sqrt{1+x^2} + x\right| \Big|_0^4$$

$$= \left(2\sqrt{17} + \tfrac{1}{2}\ln\left|\sqrt{17} + 4\right|\right) - \tfrac{1}{2}\ln|1|$$

$$\approx 2\sqrt{17} + \tfrac{1}{2}\ln\left|\sqrt{17} + 4\right|$$

$$\approx 8.2462 + 1.04735$$

$$\approx 9.294.$$

Figure 9.4 shows the graph of $y = 0.5x^2$ over the interval $x \in [0, 4]$, where the length of the straight line joining $(0, 0)$ and $(4, 8)$ is $\sqrt{80} \approx 8.94$, which provides a minimum value for the arc length. And by reducing the parabola to a chain of straight-line segments whose $\Delta x = 0.25$, the arc length equals 9.291, which confirms the accuracy of the above answer.

Before moving on, here is an alternative solution to the original integral

$$\int \sqrt{1+x^2}\, dx.$$

To remove the radical we let $x = \sinh\theta$ where $dx/d\theta = \cosh\theta$ and continue with an indefinite integral. Therefore,

$$s = \int \sqrt{1 + \sinh^2\theta} \cdot \cosh\theta\, d\theta$$

$$= \int \sqrt{\cosh^2\theta} \cdot \cosh\theta\, d\theta$$

$$= \int \cosh^2\theta\, d\theta.$$

But $2\cosh^2\theta = \cosh(2\theta) + 1$, therefore,

$$s = \frac{1}{2}\int \cosh(2\theta) + 1 \; d\theta$$
$$= \frac{1}{2}\int \cosh(2\theta) + \frac{1}{2}\int 1 \; d\theta$$
$$= \frac{1}{4}\sinh(2\theta) + \frac{1}{2}\theta + C.$$

But $\sinh(2\theta) = 2\cosh\theta \cdot \sinh\theta$, therefore,

$$s = \frac{1}{2}\cosh\theta \cdot \sinh\theta + \frac{1}{2}\theta + C. \tag{9.3}$$

Apart from the constant C, (9.3) contains two parts. The first part is transformed back to the original independent variable x by substituting $\sinh\theta = x$ and $\cosh\theta = \sqrt{1+x^2}$:

$$\frac{1}{2}\cosh\theta \cdot \sinh\theta = \frac{1}{2}x\sqrt{1+x^2}.$$

The second part is transformed back to the original independent variable x as follows:

$$x = \sinh\theta$$
$$= \frac{1}{2}\left(e^\theta - e^{-\theta}\right)$$
$$2x = e^\theta - e^{-\theta}$$
$$2xe^\theta = \left(e^\theta\right)^2 - 1$$
$$\left(e^\theta\right)^2 - 2xe^\theta - 1 = 0$$

which is a quadratic in e^θ, where $a = 1, \; b = -2x, \; c = -1$. Therefore,

$$e^\theta = \frac{-b \pm \sqrt{b^2 - 4ac}}{2a}$$
$$= \frac{2x \pm \sqrt{4x^2 + 4}}{2}$$
$$= x \pm \sqrt{1+x^2}.$$

However, as $e^\theta > 1$, therefore,

$$e^\theta = x + \sqrt{1+x^2}$$
$$\theta = \ln\left|x + \sqrt{1+x^2}\right|$$
$$\frac{1}{2}\theta = \frac{1}{2}\ln\left|x + \sqrt{1+x^2}\right|.$$

Combining these two parts together, and introducing a definite integral, we have

$$\int \sqrt{1+x^2}\, dx = \tfrac{1}{2}x\sqrt{1+x^2} + \tfrac{1}{2}\ln\left|\sqrt{1+x^2}+x\right| \tag{9.4}$$

which agrees with the first result.

Using the same technique, one can show that

$$\int \sqrt{x^2+a^2} = \tfrac{1}{2}x\sqrt{x^2+a^2} + \tfrac{1}{2}a^2\ln\left|x+\sqrt{x^2+a^2}\right| + C \tag{9.5}$$

$$\int \sqrt{x^2-a^2} = \tfrac{1}{2}x\sqrt{x^2+a^2} - \tfrac{1}{2}a^2\ln\left|x+\sqrt{x^2+a^2}\right| + C. \tag{9.6}$$

9.3.4 Arc Length of $y = x^{\frac{3}{2}}$

Let's find the length of the curve $y = x^{\frac{3}{2}}$ over the interval $x \in [0, 4]$.

$$\frac{dy}{dx} = \tfrac{3}{2}x^{\frac{1}{2}}$$

therefore,

$$s = \int_0^4 \sqrt{1+\left(\frac{dy}{dx}\right)^2}\, dx$$

$$= \int_0^4 \sqrt{1+\tfrac{9}{4}x}\, dx$$

$$= \int_0^4 \left(1+\tfrac{9}{4}x\right)^{\frac{1}{2}}\, dx$$

Let $u = 1 + \tfrac{9}{4}x$, then $dx = \tfrac{4}{9}\, du$.
The limits for u are:

$$x = 0, \quad u = 1,$$
$$x = 4, \quad u = 10,$$

$$s = \tfrac{4}{9}\int_1^{10} u^{\frac{1}{2}}\, du$$

$$= \tfrac{4}{9}\cdot\tfrac{2}{3}u^{\frac{3}{2}}\,\Big|_1^{10}$$

$$\approx \tfrac{8}{27}(31.62277 - 1)$$

$$\approx 9.07.$$

9.3.5 Arc Length of a Sine Curve

A radical inside the integrand does present problems, and often makes it difficult to integrate the expression. For example, consider the apparently, simple task of finding the arc length of $y = \sin x$ over the interval $x \in [0, 2\pi]$.

$$\frac{dy}{dx} = \cos x$$

therefore,

$$s = \int_0^{2\pi} \sqrt{1 + \left(\frac{dy}{dx}\right)^2} \, dx$$

$$= \int_0^{2\pi} \sqrt{1 + \cos^2 x} \, dx.$$

At this point, we have a problem, as it is not obvious how to integrate $\sqrt{1 + \cos^2 x}$. It is what is called an *elliptic integral* of the second kind, and beyond the remit of this introductory book. Dividing the sine wave into a series of line segments, and using the theorem of Pythagoras, we discover that the length converges as follows:

$$10° \text{ steps} \approx 7.6373564$$
$$5° \text{ steps} \approx 7.6396352$$
$$2° \text{ steps} \approx 7.6402736$$
$$1° \text{ steps} \approx 7.6403648.$$

9.3.6 Arc Length of a Hyperbolic Cosine Function

Finding the arc length of $y = 3\cosh\left(\frac{1}{3}x\right)$ over the interval $x \in [-3, 3]$ turns out to be much easier than $y = \sin x$:

$$\frac{dy}{dx} = \sinh\left(\frac{1}{3}x\right)$$

therefore,

$$s = \int_{-3}^{3} \sqrt{1 + \left(\frac{dy}{dx}\right)^2} \, dx$$

$$= \int_{-3}^{3} \sqrt{1 + \sinh^2\left(\frac{1}{3}x\right)} \, dx$$

Fig. 9.5 The graph of $y = 3 \cosh\left(\frac{1}{3}x\right)$

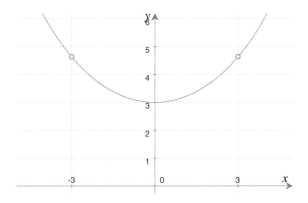

$$= \int_{-3}^{3} \sqrt{\cosh^2\left(\frac{1}{3}x\right)} \, dx$$

$$= \int_{-3}^{3} \cosh\left(\frac{1}{3}x\right) \, dx$$

$$= 3 \sinh\left(\frac{1}{3}x\right) \Big|_{-3}^{3}$$

$$= 3[\sinh 1 - \sinh(-1)]$$

$$= 3\left[\frac{e^1 - e^{-1}}{2} - \frac{e^{-1} - e^1}{2}\right]$$

$$= 3\left(e^1 - e^{-1}\right)$$

$$\approx 7.05.$$

Figure 9.5 shows the graph of $y = 3 \cosh\left(\frac{1}{3}x\right)$.

9.3.7 Arc Length of Parametric Functions

Parametric functions take the form

$$x = f_x(t)$$
$$y = f_y(t)$$

where f_x and f_y are independent functions, but share a common parameter t. In order to compute the arc length of such a function we need to derive the derivative dy/dx. The individual derivatives are dx/dt and dy/dt and can be combined to produce dy/dx as follows

$$\frac{dy}{dx} = \frac{dy/dt}{dx/dt}$$

which means that (9.2) can be written as

$$s = \int_a^b \sqrt{1 + \left(\frac{dy/dt}{dx/dt}\right)^2} \, dx$$

$$= \int_a^b \sqrt{\frac{(dx/dt)^2 + (dy/dt)^2}{(dx/dt)^2}} \, dx$$

$$= \int_a^b \sqrt{\left(\frac{dx}{dt}\right)^2 + \left(\frac{dy}{dt}\right)^2} \frac{dt}{dx} \, dx$$

$$s = \int_a^b \sqrt{\left(\frac{dx}{dt}\right)^2 + \left(\frac{dy}{dt}\right)^2} \, dt. \tag{9.7}$$

A similar analysis can be performed for 3D parametric curves, where we have

$$x = f_x(t)$$
$$y = f_y(t)$$
$$z = f_z(t)$$

and

$$s = \int_a^b \sqrt{\left(\frac{dx}{dt}\right)^2 + \left(\frac{dy}{dt}\right)^2 + \left(\frac{dz}{dt}\right)^2} \, dt. \tag{9.8}$$

Writing a parametric function as a Cartesian vector:

$$\mathbf{r}(t) = x(t)\mathbf{i} + y(t)\mathbf{j} + z(t)\mathbf{k}$$

its derivative is

$$\mathbf{r}'(t) = x'(t)\mathbf{i} + y'(t)\mathbf{j} + z'(t)\mathbf{k}.$$

The derivative, $\mathbf{r}'(t)$ is the tangent vector to the curve, whose magnitude is $||\mathbf{r}'(t)||$, therefore,

$$||\mathbf{r}'(t)|| = \sqrt{\left(x'(t)\right)^2 + \left(y'(t)\right)^2 + \left(z'(t)\right)^2}. \tag{9.9}$$

We can use (9.7) and (9.8) to solve various problems in two and three dimensions.

9.3.8   Arc Length of a Circle

Let's start with the parametric equation of a circle with radius r, by computing the arc length of one quadrant, and multiplying this by 4:

$$\mathbf{r}(t) = x(t)\mathbf{i} + y(t)\mathbf{j}, \quad t \in [0, 2\pi]$$
$$x(t) = r \cos t$$
$$y(t) = r \sin t.$$

Differentiating $x(t)$ and $y(t)$:

$$\frac{dx}{dt} = -r \sin t$$

$$\frac{dy}{dt} = r \cos t$$

and substituting them in (9.7):

$$\begin{aligned}
s &= 4 \int_0^{\pi/2} \sqrt{\left(-r \sin t\right)^2 + \left(r \cos t\right)^2}\, dt \\
&= 4r \int_0^{\pi/2} \sqrt{\sin^2 t + \cos^2 t}\, dt \\
&= 4r \int_0^{\pi/2} 1\, dt \\
&= 4rt \,\Big|_0^{\pi/2} \\
&= 2\pi r.
\end{aligned}$$

9.3.9 Arc Length of an Ellipse

Let's follow the circle with an ellipse, which is represented parametrically:

$$\mathbf{r}(t) = x(t)\mathbf{i} + y(t)\mathbf{j}, \quad t \in [0, 2\pi]$$
$$x(t) = a \cos t$$
$$y(t) = b \sin t.$$

Differentiating $x(t)$ and $y(t)$:

$$\frac{dx}{dt} = -a \sin t$$

$$\frac{dy}{dt} = b \cos t$$

and substituting them in (9.7):

$$s = 4 \int_0^{\pi/2} \sqrt{\left(-a \sin t\right)^2 + \left(b \cos t\right)^2} \, dt$$

$$= 4 \int_0^{\pi/2} \sqrt{a^2 \sin^2 t + b^2 \cos^2 t} \, dt$$

$$= 4 \int_0^{\pi/2} \sqrt{a^2(1 - \cos^2 t) + b^2 \cos^2 t} \, dt$$

$$= 4 \int_0^{\pi/2} \sqrt{a^2 - (a^2 - b^2) \cos^2 t} \, dt$$

$$= 4a \int_0^{\pi/2} \sqrt{1 - \epsilon^2 \cos^2 t} \, dt \tag{9.10}$$

where $\epsilon = \sqrt{1 - b^2/a^2}$ is the eccentricity of the ellipse. Equation (9.10) is an *elliptic integral*, and can only be solved numerically, as no standard function is available. However, using the binomial theorem, and cosine integrals (www.pages.pacificcoast. net/~cazelais/250a/ellipse-length.pdf), it can be shown that

$$s = 2\pi a \left(1 - \left(\frac{1}{2}\right)^2 \frac{\epsilon^2}{1} - \left(\frac{1 \cdot 3}{2 \cdot 4}\right)^2 \frac{\epsilon^4}{3} - \left(\frac{1 \cdot 3 \cdot 5}{2 \cdot 4 \cdot 6}\right)^2 \frac{\epsilon^6}{5} - \left(\frac{1 \cdot 3 \cdot 5 \cdot 7}{2 \cdot 4 \cdot 6 \cdot 8}\right)^2 \frac{\epsilon^8}{7} - \cdots\right).$$
$$\tag{9.11}$$

Given an ellipse where $a = 5$ and $b = 4$, then $\epsilon = 0.6$. Let's compute (9.11) by including increasing number of terms:

$$s \approx 2\pi a \left(1 - \left(\frac{1}{2}\right)^2 \frac{\epsilon^2}{1}\right) \approx 28.58849$$

$$s \approx 2\pi a \left(1 - \left(\frac{1}{2}\right)^2 \frac{\epsilon^2}{1} - \left(\frac{1 \cdot 3}{2 \cdot 4}\right)^2 \frac{\epsilon^4}{3}\right) \approx 28.58242$$

$$s \approx 2\pi a \left(1 - \left(\frac{1}{2}\right)^2 \frac{\epsilon^2}{1} - \left(\frac{1 \cdot 3}{2 \cdot 4}\right)^2 \frac{\epsilon^4}{3} - \left(\frac{1 \cdot 3 \cdot 5}{2 \cdot 4 \cdot 6}\right)^2 \frac{\epsilon^6}{5}\right) \approx 28.36901$$

$$s \approx 2\pi a \left(1 - \left(\frac{1}{2}\right)^2 \frac{\epsilon^2}{1} - \left(\frac{1 \cdot 3}{2 \cdot 4}\right)^2 \frac{\epsilon^4}{3} - \left(\frac{1 \cdot 3 \cdot 5}{2 \cdot 4 \cdot 6}\right)^2 \frac{\epsilon^6}{5} - \left(\frac{1 \cdot 3 \cdot 5 \cdot 7}{2 \cdot 4 \cdot 6 \cdot 8}\right)^2 \frac{\epsilon^8}{7}\right) \approx 28.36338$$

therefore, the ellipse's perimeter is approximately 28.36.

9.3.10 Arc Length of a Helix

A constant-pitch helix is shown in Fig. 9.6, and can be defined as

Fig. 9.6 A constant-pitch helix

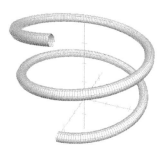

$$\mathbf{r}(t) = r \cos t\mathbf{i} + r \sin t\mathbf{j} + t\mathbf{k}$$

where r is the radius in the xy-plane. If $r = 2$, then

$$\mathbf{r}(t) = 2 \cos t\mathbf{i} + 2 \sin t\mathbf{j} + t\mathbf{k}.$$

Its arc length is computed using

$$\mathbf{r}'(t) = -2 \sin t\mathbf{i} + 2 \cos t\mathbf{j} + \mathbf{k}$$

where $t \in [0, 4\pi]$. Therefore, using (9.8)

$$\begin{aligned}
s &= \int_0^{4\pi} \sqrt{4 \sin^2 t + 4 \cos^2 t + 1} \; dt \\
&= \int_0^{4\pi} \sqrt{5} \; dt \\
&= \sqrt{5} t \Big|_0^{4\pi} \\
&\approx 28.1.
\end{aligned}$$

Thus the length of the helix over two turns is ≈ 28.1.

9.3.11 Arc Length of a 2D Quadratic Bézier Curve

For an introduction to Bézier curves, see my book *Mathematics for Computer Graphics*, Vince (2017).

A 2D quadratic Bézier curve is represented as:

$$\mathbf{r}(t) = \begin{bmatrix} x(t) \\ y(t) \end{bmatrix}, \quad t \in [0, \; 1]$$

$$x(t) = x_0(1 - 2t + t^2) + x_1(2t - 2t^2) + x_2 t^2$$

$$y(t) = y_0(1 - 2t + t^2) + y_1(2t - 2t^2) + y_2 t^2.$$

Fig. 9.7 A 2D Bézier curve

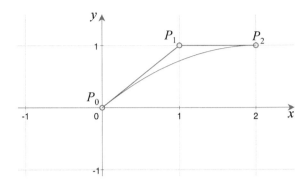

Differentiating $x(t)$ and $y(t)$:

$$\frac{dx}{dt} = x_0(2t - 2) + x_1(2 - 4t) + x_2 2t$$

$$\frac{dy}{dt} = y_0(2t - 2) + y_1(2 - 4t) + y_2 2t.$$

Let's take a simple example, with $P_0 = (0, 0)$, $P_1 = (1, 1)$ and $P_2 = (2, 1)$, as shown in Fig. 9.7. Using Pythagoras, the arc length must be slightly longer than $\sqrt{5} \approx 2.236$. Therefore,

$$\frac{dx}{dt} = 2$$

$$\frac{dy}{dt} = 2 - 2t$$

$$s = \int_0^1 \sqrt{2^2 + (2 - 2t)^2}\, dt$$

$$= \int_0^1 \sqrt{8 - 8t + 4t^2}\, dt$$

$$= 2 \int_0^1 \sqrt{t^2 - 2t + 2}\, dt$$

$$= 2 \int_0^1 \sqrt{(t - 1)^2 + 1^2}\, dt.$$

Using (9.5):

$$\int \sqrt{x^2 + a^2}\, dx = \tfrac{1}{2}x\sqrt{x^2 + a^2} + \tfrac{1}{2}a^2 \ln \left| x + \sqrt{x^2 + a^2} \right| + C$$

therefore, let $x = t - 1$ and $a = 1$

Table 9.1 Computing the line-segment lengths

t	x	y	Δx	Δy	$(\Delta x)^2$	$(\Delta y)^2$	$\sqrt{(\Delta x)^2 + (\Delta y)^2}$
0	0	0	0	0	0	0	0
0.2	0.4	0.36	0.4	0.36	0.16	0.1296	0.538145
0.4	0.8	0.64	0.4	0.28	0.16	0.0784	0.488262
0.6	1.2	0.84	0.4	0.2	0.16	0.04	0.447214
0.8	1.6	0.96	0.4	0.12	0.16	0.0144	0.417612
1	2	1	0.4	0.04	0.16	0.0016	0.401995

$$s = 2\left(\frac{t-1}{2}\sqrt{t^2 - 2t + 2} + \tfrac{1}{2}\ln\left|t - 1 + \sqrt{t^2 - 2t + 2}\right|\right)\Big|_0^1$$

$$= (t-1)\sqrt{t^2 - 2t + 2} + \ln\left|(t-1) + \sqrt{t^2 - 2t + 2}\right|\Big|_0^1$$

$$= -\sqrt{2} + \ln\left|-1 + \sqrt{2}\right|$$

$$\approx 2.296.$$

By making $P_1 = (1, 0.5)$, the Bézier curve becomes a straight line with length $\sqrt{5}$.

$$\frac{dx}{dt} = 2$$

$$\frac{dy}{dt} = 1$$

$$s = \int_0^1 \sqrt{2^2 + 1^2}\, dt$$

$$= \sqrt{5}t \,\Big|_0^1$$

$$\approx 2.236.$$

We can approximate the arc length by reducing the curve to a chain of straight-line segments, and summing their lengths. Table 9.1 shows the calculations. The sum of the right-hand column gives the total length of the line segments, which is 2.293, and is very close to the integral: 2.296.

9.3.12 Arc Length of a 3D Quadratic Bézier Curve

A 3D quadratic Bézier curve is represented as:

$$\mathbf{r}(t) = \begin{bmatrix} x(t) \\ y(t) \\ z(t) \end{bmatrix}, \quad t \in [0, 1]$$

$$x(t) = x_0(1 - 2t + t^2) + x_1(2t - 2t^2) + x_2 t^2$$
$$y(t) = y_0(1 - 2t + t^2) + y_1(2t - 2t^2) + y_2 t^2$$
$$z(t) = z_0(1 - 2t + t^2) + z_1(2t - 2t^2) + z_2 t^2.$$

Differentiating $x(t)$, $y(t)$ and $z(t)$:

$$\frac{dx}{dt} = x_1(2t - 2) + x_2(2 - 4t) + x_3 2t$$
$$\frac{dy}{dt} = y_1(2t - 2) + y_2(2 - 4t) + y_3 2t$$
$$\frac{dz}{dt} = z_1(2t - 2) + z_2(2 - 4t) + z_3 2t.$$

Let's take a simple example, with $P_0 = (0, 0, 0)$, $P_1 = (1, 1, 1)$ and $P_2 = (2, 1, 1)$. Therefore,

$$\frac{dx}{dt} = 2$$
$$\frac{dy}{dt} = 2 - 2t$$
$$\frac{dz}{dt} = 2 - 2t$$

$$s = \int_0^1 \sqrt{(2)^2 + (2 - 2t)^2 + (2 - 2t)^2} \, dt$$
$$= \int_0^1 \sqrt{8 - 8t + 4t^2 + 4 - 8t + 4t^2} \, dt$$
$$= \int_0^1 \sqrt{8t^2 - 16t + 12} \, dt$$
$$= 2\sqrt{2} \int_0^1 \sqrt{t^2 - 2t + 1.5} \, dt$$
$$= 2\sqrt{2} \int_0^1 \sqrt{(t - 1)^2 + \left(1/\sqrt{2}\right)^2} \, dt.$$

Using (9.5):

$$\int \sqrt{x^2 + a^2}\, dx = \tfrac{1}{2}x\sqrt{x^2 + a^2} + \tfrac{1}{2}a^2 \ln\left|x + \sqrt{x^2 + a^2}\right| + C$$

therefore, let $x = t - 1$ and $a = 1/\sqrt{2}$

$$
\begin{aligned}
s &= 2\sqrt{2}\left(\frac{t-1}{2}\sqrt{t^2 - 2t + 1.5} + \tfrac{1}{2 \cdot 2} \ln\left|t - 1 + \sqrt{t^2 - 2t + 1.5}\right|\right)\Bigg|_0^1 \\
&= 2\sqrt{2}\left(\left(\tfrac{1}{4}\ln\left|\sqrt{0.5}\right|\right) - \left(\tfrac{-1}{2}\sqrt{1.5} + \tfrac{1}{4}\ln\left|-1 + \sqrt{1.5}\right|\right)\right) \\
&\approx 2\sqrt{2}\left(-0.086643 - (-0.612372 - 0.373197)\right) \\
&\approx 2\sqrt{2}\left(-0.086643 + 0.837117\right) \\
&\approx 2\sqrt{2} \cdot 0.92376 \\
&\approx 2.612788.
\end{aligned}
$$

9.3.13 Arc Length Parameterisation of a 3D Line

One useful tool in computer animation is the ability to move along a 3D curve in a controlled manner. Unfortunately, this turns out to be a difficult calculation, as it involves integrating a function within a radical, and very often there is no standard solution. This means employing some high-level mathematics to secure an approximate numerical solution.

In this section we examine the arc-length parameterisation of a straight line, and in the following section for a parametric helix curve.

A vector-valued function normally takes the form:

$$
\begin{aligned}
\mathbf{r}(t) &= x(t)\mathbf{i} + y(t)\mathbf{j} + z(t)\mathbf{k} \\
\mathbf{r}'(t) &= x'(t)\mathbf{i} + y'(t)\mathbf{j} + z'(t)\mathbf{k} \\
\|\mathbf{r}'(t)\| &= \sqrt{\left(x'(t)\right)^2 + \left(y'(t)\right)^2 + \left(z'(t)\right)^2}
\end{aligned}
$$

therefore, we can define an arc length function as

$$s(t) = \int_a^t \|\mathbf{r}'(u)\|\, du = \int_a^t \sqrt{\left(x'(t)\right)^2 + \left(y'(t)\right)^2 + \left(z'(t)\right)^2}\, du$$

and the fundamental theorem of Calculus states (https://en.wikipedia.org/wiki/Fundamental_theorem_of_Calculus), if $f(u)$ is well behaved, and

$$s(t) = \int_a^t f(u)\, du$$

then

$$\frac{ds}{dt} = f(t),$$

therefore,

$$\frac{ds}{dt} = ||\mathbf{r}'(t)||.$$

For example, (9.12) describes a 2D straight line 5 units long:

$$\mathbf{r}(t) = 4t\mathbf{i} + 3t\mathbf{j}, \quad t \in [0, \ 1] \tag{9.12}$$

where any point is determined by the value of t. When $t = 1$, the line's length is 5, and when $t = 0.5$, the line's length is 2.5. Clearly, the line's length s is given by $s = 5t$. The parameter t, which could stand for time, but is probably just an independent parameter. Now say we wish to write (9.12) using the line's length s, then we must change every occurrence of t for s:

$$\mathbf{r}(s) = \tfrac{4}{5}s\mathbf{i} + \tfrac{3}{5}s\mathbf{j}, \quad s \in [0, \ t].$$

Knowing that $s = 5t$, means that $t = s/5$. We can now locate points anywhere along the line as a proportion of the line's length s.

If (9.12) is written generally:

$$\mathbf{r}(t) = at\mathbf{i} + bt\mathbf{j}, \quad t \in [0, \ 1]$$

the line's length is $\sqrt{a^2 + b^2}$, and any distance along the line is given by

$$s = t\sqrt{a^2 + b^2}$$

therefore $t = s/\sqrt{a^2 + b^2}$, and

$$\mathbf{r}(s) = \frac{as}{\sqrt{a^2 + b^2}}\mathbf{i} + \frac{bs}{\sqrt{a^2 + b^2}}\mathbf{j}.$$

For a 3D line

$$\mathbf{r}(t) = \begin{bmatrix} x(t) \\ y(t) \\ z(t) \end{bmatrix}, \quad t \in [0, \ 1]$$

$$x(t) = at$$
$$y(t) = bt$$
$$z(t) = ct$$

$$\mathbf{r}(t(s)) = \begin{bmatrix} \dfrac{as}{\sqrt{a^2 + b^2 + c^2}} \\ \dfrac{bs}{\sqrt{a^2 + b^2 + c^2}} \\ \dfrac{cs}{\sqrt{a^2 + b^2 + c^2}} \end{bmatrix}.$$

The above reasoning seems straight forward, but we must find a strategy using Calculus, so that we can parameterise 3D curves in terms of their arc length. So let's use Calculus to parameterise a 3D straight line.

Starting with the vector-valued function (9.13):

$$\mathbf{r}(t) = \begin{bmatrix} x(t) \\ y(t) \\ z(t) \end{bmatrix}, \quad t \in [0, \ 1] \tag{9.13}$$

$$x(t) = at$$
$$y(t) = bt$$
$$z(t) = ct.$$

Differentiate $\mathbf{r}(t)$

$$\mathbf{r}'(t) = \begin{bmatrix} a \\ b \\ c \end{bmatrix}.$$

The length of $\mathbf{r}'(t)$ is

$$||\mathbf{r}'(t)|| = \sqrt{a^2 + b^2 + c^2}.$$

We already know that

$$s(t) = \int ||\mathbf{r}'(t)|| \ dt$$

but this integrates with respect to t, and we want to integrate the arc length over the interval $[0, \ t]$. Therefore, we use another parameter, say u, such that

$$s(t) = \int_0^t ||\mathbf{r}'(u)|| \ du.$$

Therefore, we can write

$$s(t) = \int_0^t ||\mathbf{r}'(u)|| \ du$$
$$= \int_0^t \sqrt{a^2 + b^2 + c^2} \ du$$

$$= u\sqrt{a^2 + b^2 + c^2}\ \Big|_0^t$$

$$= t\sqrt{a^2 + b^2 + c^2}.$$

So $s = t\sqrt{a^2 + b^2 + c^2}$ and $t = s/\sqrt{a^2 + b^2 + c^2}$. Substituting t back in (9.13):

$$\mathbf{r}(t(s)) = \begin{bmatrix} \dfrac{as}{\sqrt{a^2 + b^2 + c^2}} \\[2mm] \dfrac{bs}{\sqrt{a^2 + b^2 + c^2}} \\[2mm] \dfrac{cs}{\sqrt{a^2 + b^2 + c^2}} \end{bmatrix}. \tag{9.14}$$

With the limits of t being $[0, 1]$, the limits of s in (9.14) are $\left[0,\ t\sqrt{a^2 + b^2 + c^2}\ \right]$.

9.3.14   Arc Length Parameterisation of a Helix

Now let's apply the above reasoning to a helix curve, which is chosen to keep the maths simple. Starting with the vector-valued function

$$\mathbf{r}(t) = \begin{bmatrix} a\cos t \\ a\sin t \\ bt \end{bmatrix}, \quad t \in [0,\ 2\pi] \tag{9.15}$$

and differentiating:

$$\mathbf{r}'(t) = \begin{bmatrix} -a\sin t \\ a\cos t \\ b \end{bmatrix}.$$

The length of the tangent vector $\mathbf{r}'(t)$ is

$$\|\mathbf{r}'(t)\| = \sqrt{a^2 \sin^2 t + a^2 \cos^2 t + b^2}$$

$$= \sqrt{a^2\left(\sin^2 t + \cos^2 t\right) + b^2}$$

$$= \sqrt{a^2 + b^2}.$$

The arc length of the helix over the interval $[0,\ t]$ is

$$s(t) = \int_o^t \|\mathbf{r}'(u)\|\ du$$

$$= \int_o^t \sqrt{a^2 + b^2}\ du$$

$$= u\sqrt{a^2 + b^2} \ \Big|_0^t$$

$$= t\sqrt{a^2 + b^2}.$$

Therefore, $t = s/\sqrt{a^2 + b^2}$, which substituted in (9.15) gives:

$$\mathbf{r}(t(s)) = \begin{bmatrix} a\cos\left(\dfrac{s}{\sqrt{a^2 + b^2}}\right) \\ a\sin\left(\dfrac{s}{\sqrt{a^2 + b^2}}\right) \\ \dfrac{bs}{\sqrt{a^2 + b^2}} \end{bmatrix}, \quad s \in \left[0, \ t\sqrt{a^2 + b^2}\right]. \tag{9.16}$$

Let's illustrate (9.16) with an example.

Given $a = 4$, $b = 3$, and $t \in [0, 2\pi]$, then $s \in [0, \ 5t]$

$$\mathbf{r}(s) = \begin{bmatrix} 4\cos(s/5) \\ 4\sin(s/5) \\ 3s/5 \end{bmatrix}$$

$$\mathbf{r}(0) = \begin{bmatrix} 4\cos(0) \\ 4\sin(0) \\ 0 \end{bmatrix} = \begin{bmatrix} 4 \\ 0 \\ 0 \end{bmatrix}$$

$$\mathbf{r}(10\pi) = \begin{bmatrix} 4\cos(2\pi) \\ 4\sin(2\pi) \\ 6\pi \end{bmatrix} = \begin{bmatrix} 4 \\ 0 \\ 6\pi \end{bmatrix}.$$

9.3.15 Positioning Points on a Straight Line Using a Square Law

One can position points on a line or a fixed-pitch helix using their parametric equation. Nevertheless, I will show this process in terms of arc length. To illustrate this, consider the original 2D line equation:

$$\mathbf{r}(s) = \begin{bmatrix} \frac{4}{5}s \\ \frac{3}{5}s \end{bmatrix}, \quad s \in [0, \ 5].$$

Table 9.2 The value of $\mathbf{r}(s)$
for different values of s

s	$(s/5)^2$	$\mathbf{r}(s)$
0	0	[0, 0]
0.5	0.01	[0.04, 0.03]
1	0.04	[0.16, 0.12]
1.5	0.09	[0.36, 0.27]
2	0.16	[0.64, 0.48]
2.5	0.25	[1, 0.75]
3	0.36	[1.44, 1.08]
3.5	0.49	[1.96, 1.47]
4	0.64	[2.56, 1.92]
4.5	0.81	[3.24, 2.43]
5	1.0	[5, 3]

Fig. 9.8 Points located
along a line with a
square-law distribution

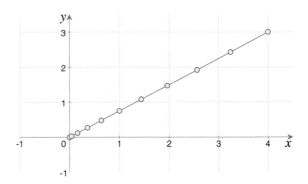

Rewriting this as

$$\mathbf{r}(s) = \begin{bmatrix} 4\left(\dfrac{s}{5}\right)^2 \\ 3\left(\dfrac{s}{5}\right)^2 \end{bmatrix}, \quad s \in [0,\ 5].$$

locates points along the line with a square-law distribution. Table 9.2 shows the values
of $\mathbf{r}(s)$ for different values of s, and Fig. 9.8 shows the points located along the line.

9.3.16   Positioning Points on a Helix Curve Using a Square Law

Now let's arrange points along a helix using a square law. We start with the vector-
valued function for a helix

$$\mathbf{r}(t) = \begin{bmatrix} 2\cos t \\ 2\sin t \\ t \end{bmatrix}, \quad t \in [0,\ 2\pi]$$

which becomes

$$\mathbf{r}(s) = \begin{bmatrix} 2\cos\left(s/\sqrt{5}\right) \\ 2\sin\left(s/\sqrt{5}\right) \\ s/\sqrt{5} \end{bmatrix}, \quad s \in \left[0,\ 2\pi\sqrt{5}\right]$$

$$= \begin{bmatrix} 2\cos\left(2\pi s/\sqrt{5}\right) \\ 2\sin\left(2\pi s/\sqrt{5}\right) \\ 2\pi s/\sqrt{5} \end{bmatrix}, \quad s \in \left[0,\ \sqrt{5}\right] \tag{9.17}$$

$$= \begin{bmatrix} 2\cos\left(2\pi s^2/5\right) \\ 2\sin\left(2\pi s^2/5\right) \\ 2\pi s^2/5 \end{bmatrix}, \quad s \in \left[0,\ \sqrt{5}\right]. \tag{9.18}$$

Equation (9.17) locates points along the helix as a linear function of the arc length, but by squaring $s/\sqrt{5}$, we obtain a square-law distribution (9.18). Table 9.3 shows the necessary calculations, and Fig. 9.9 shows the helix, looking down the z-axis.

Naturally, we could have used any type of law to distribute the points. This is not the problem. The real problem is securing a vector-valued function for the arc-length parameterisation. We have already seen the difficulty in computing the arc length of a Bézier curve, and cubic B-splines are equally obscure.

Table 9.3 The value of $\mathbf{r}(s)$ for different values of s

s	$s^2/5$	$2\pi s^2/5$	$\mathbf{r}(s)$
0	0	0	[2, 0, 0]
0.2236	0.01	0.063	[1.996, 0.126, 0.063]
0.4472	0.04	0.251	[1.937, 0.497, 0.251]
0.6708	0.09	0.565	[1.689, 1.071, 0.565]
0.8944	0.16	1.005	[1.072, 1.688, 1.005]
1.118	0.25	1.571	[0, 2, 1.571]
1.3416	0.36	2.262	[−1.275, 1.541, 2.262]
1.5652	0.49	3.079	[−1.996, 0.125, 3.079]
1.7888	0.64	4.021	[−1.275, −1.541, 4.021]
2.0124	0.81	5.089	[0.736, −1.859, 5.089]
2.236	1.0	6.283	[2, 0, 6.283]

Fig. 9.9 Points located
along a helix with a
square-law distribution

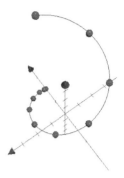

9.3.17   Arc Length Using Polar Coordinates

Polar coordinates are sometimes more convenient than Cartesian coordinates when
describing functions involving trigonometric functions. For example, Fig. 9.10 shows
the correspondence between a point (x, y) and its polar coordinates (r, θ), where

$$x = r \cos \theta$$
$$y = r \sin \theta$$

and as $r = f(\theta)$, we have the product of two functions. Rewriting (9.7) in terms of
θ we have

$$s = \int_{\theta_1}^{\theta_2} \sqrt{\left(\frac{dx}{d\theta}\right)^2 + \left(\frac{dy}{d\theta}\right)^2}\, d\theta. \tag{9.19}$$

To find $dx/d\theta$ and $dy/d\theta$ we have to employ the product rule:

$$x = u(\theta) \cdot v(\theta)$$
$$\frac{dx}{d\theta} = u(\theta)\frac{dv}{d\theta} + v(\theta)\frac{du}{d\theta}$$

Fig. 9.10 The
correspondence between
Cartesian and polar
coordinates

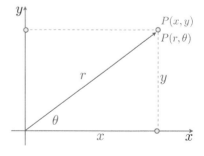

therefore,

$$x = r\cos\theta$$

$$\frac{dx}{d\theta} = -r\sin\theta + \frac{dr}{d\theta}\cos\theta \tag{9.20}$$

$$y = r\sin\theta$$

$$\frac{dy}{d\theta} = r\cos\theta + \frac{dr}{d\theta}\sin\theta \tag{9.21}$$

substituting (9.20) and (9.21) in (9.19):

$$s = \int_{\theta_1}^{\theta_2}\sqrt{\left(-r\sin\theta + \frac{dr}{d\theta}\cos\theta\right)^2 + \left(r\cos\theta + \frac{dr}{d\theta}\sin\theta\right)^2}\,d\theta$$

$$= \int_{\theta_1}^{\theta_2}\sqrt{r^2\sin^2\theta + \left(\frac{dr}{d\theta}\right)^2\cos^2\theta + r^2\cos^2\theta + \left(\frac{dr}{d\theta}\right)^2\sin^2\theta}\,d\theta$$

$$= \int_{\theta_1}^{\theta_2}\sqrt{r^2 + \left(\frac{dr}{d\theta}\right)^2}\,d\theta$$

therefore, the arc length is

$$s = \int_{\theta_1}^{\theta_2}\sqrt{r^2 + \left(\frac{dr}{d\theta}\right)^2}\,d\theta. \tag{9.22}$$

Let's test (9.22) with the arc length of a circle, where $r = 2$ and $\theta \in [0, 2\pi]$. Therefore,

$$\frac{dr}{d\theta} = 0$$

and

$$s = \int_0^{2\pi}\sqrt{2^2 + 0^2}\,d\theta$$

$$= 2\int_0^{2\pi} 1\,d\theta$$

$$= 2\theta\,\Big|_0^{2\pi}$$

$$= 4\pi$$

which is very compact, as well as correct. Now let's compute the length of a logarithmic spiral.

Figure 9.11 shows the graph of a logarithmic spiral $r = 2e^{0.2\theta}$ where $\theta \in [0, 2\pi]$, whose length is calculated using (9.22) as follows.

Fig. 9.11 Polar graph of $r = 2e^{0.2\theta}$

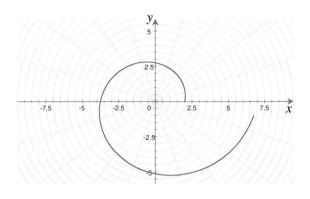

$$r = 2e^{0.2\theta}$$

$$\frac{dr}{d\theta} = 0.4e^{0.2\theta}$$

$$s = \int_0^{2\pi} \sqrt{r^2 + \left(\frac{dr}{d\theta}\right)^2} \, d\theta$$

$$= \int_0^{2\pi} \sqrt{\left(2e^{0.2\theta}\right)^2 + \left(0.4e^{0.2\theta}\right)^2} \, d\theta$$

$$= \int_0^{2\pi} \sqrt{4e^{0.4\theta} + 0.16e^{0.4\theta}} \, d\theta$$

$$= \int_0^{2\pi} \sqrt{4.16e^{0.4\theta}} \, d\theta$$

$$= \sqrt{4.16} \int_0^{2\pi} e^{0.2\theta} \, d\theta$$

$$= \frac{\sqrt{4.16}}{0.2} \cdot e^{0.2\theta} \Big|_0^{2\pi}$$

$$= \frac{\sqrt{4.16}}{0.2} \left(e^{0.4\pi} - e^0\right)$$

$$\approx \frac{\sqrt{4.16}}{0.2} (3.5136 - 1)$$

$$\approx 25.634.$$

9.4   Summary

In this chapter we have computed the arc length of various functions using integration. However, all of the integrands contain a radical, which often makes integration difficult, if not impossible, without resorting to numerical techniques or employing software solutions.

9.4.1 Summary of Formulae

Explicit Functions

$$y = f(x)$$

$$s = \int_a^b \sqrt{1 + \left(\frac{dy}{dx}\right)^2}\, dx.$$

2D Parametric Functions

$$\mathbf{r}(t) = \begin{bmatrix} x(t) \\ y(t) \end{bmatrix}$$

$$s = \int_a^b \sqrt{\left(\frac{dx}{dt}\right)^2 + \left(\frac{dy}{dt}\right)^2}\, dt.$$

3D Parametric Functions

$$\mathbf{r}(t) = \begin{bmatrix} x(t) \\ y(t) \\ z(t) \end{bmatrix}$$

$$s = \int_a^b \sqrt{\left(\frac{dx}{dt}\right)^2 + \left(\frac{dy}{dt}\right)^2 + \left(\frac{dz}{dt}\right)^2}\, dt.$$

Polar Coordinates

$$r = f(\theta)$$

$$s = \int_{\theta_1}^{\theta_2} \sqrt{r^2 + \left(\frac{dr}{d\theta}\right)^2}\, d\theta.$$

References

www.pages.pacificcoast.net/~cazelais/250a/ellipse-length.pdf.

Vince, J. A. (2017). *Mathematics for computer graphics* (5th ed.). Berlin: Springer.

https://en.wikipedia.org/wiki/Fundamental_theorem_of_Calculus.

Chapter 10
Surface Area

10.1 Introduction

In Chap. 8 I showed how to compute the area under a graph using integration, and in this chapter I describe how single and double integration are used to compute surface areas and regions bounded by functions. Also in this chapter, we come across Jacobians, which are used to convert an integral from one coordinate system to another. To start, let's examine surfaces of revolution.

10.2 Surface of Revolution

A surface of revolution is a popular 3D modelling technique used in computer graphics for creating objects such as wine glasses and vases, where a contour is rotated about an axis. Integral Calculus provides a way to compute the area of such surfaces using

$$S = 2\pi \int_a^b f(x) \sqrt{1 + \left(\frac{dy}{dx}\right)^2} \, dx \tag{10.1}$$

where $y = f(x)$ and is differentiable over the interval $x \in [a, b]$.

To derive (10.1), consider the scenario shown in Fig. 10.1, where points P and Q are on a continuous curve generated by the function $y = f(x)$. The curve over the interval $x \in [a, b]$ is to be rotated $360°$ about the x-axis.

The coordinates of P and Q are (x_i, y_i) and (x_{i+1}, y_{i+1}) respectively, $\Delta x_i = x_{i+1} - x_i$, and Δs_i approximate to the arc length between P and Q:

$$\Delta s_i \approx \sqrt{1 + (f'(c))^2} \, \Delta x_i$$

where c is some $x \in [a, b]$ satisfying Lagrange's mean-value theorem.

© Springer Nature Switzerland AG 2019
J. Vince, *Calculus for Computer Graphics*,
https://doi.org/10.1007/978-3-030-11376-6_10

Fig. 10.1 The geometry to
create a surface of revolution

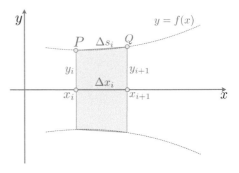

To compute the area ΔS_i swept out by the line segment PQ when rotated $360°$ about the x-axis, we use the mean radius r_i:

$$r_i = \frac{y_{i+1} + y_i}{2}$$

such that

$$\Delta S_i \approx 2\pi r_i \Delta s_i$$

$$\approx 2\pi \left(\frac{y_{i+1} + y_i}{2} \right) \sqrt{1 + (f'(c))^2} \Delta x_i.$$

As $\Delta x_i \to 0$, $y_{i+1} \approx y_i \approx f(c)$, therefore

$$\Delta S_i \approx 2\pi f(c) \sqrt{1 + (f'(c))^2} \Delta x_i.$$

Consequently, the total area swept by the arc about the x-axis is

$$S = \lim_{n \to \infty} \sum_{i=1}^{n} 2\pi f(c) \sqrt{1 + (f'(c))^2} \Delta x_i$$

$$S = 2\pi \int_a^b f(x) \sqrt{1 + \left(\frac{dy}{dx} \right)^2} \, dx. \tag{10.2}$$

Similarly, the total area swept by the arc about the y-axis is

$$S = 2\pi \int_a^b f(y) \sqrt{1 + \left(\frac{dx}{dy} \right)^2} \, dy. \tag{10.3}$$

Let's use (10.2) and (10.3) with various functions.

Fig. 10.2 Surface area of a cylinder

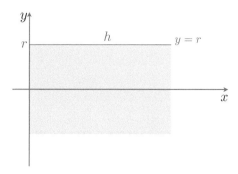

10.2.1 Surface Area of a Cylinder

To compute the surface are of a cylinder we employ the geometry shown in Fig. 10.2, where a straight horizontal line is rotated 360° about the x-axis. The function is simply $y = r$, and $x \in [0, h]$. As $y = r$, $dy/dx = 0$, and

$$S = 2\pi \int_a^b f(x) \sqrt{1 + \left(\frac{dy}{dx}\right)^2}\, dx$$

$$= 2\pi r \int_0^h 1\, dx$$

$$= 2\pi r \cdot x \,\Big|_0^h$$

$$= 2\pi r h$$

which is correct.

10.2.2 Surface Area of a Right Cone

To compute the surface area of a right cone we employ the function $y = rx/h$, where r is the cone's radius and h its height, as shown in Fig. 10.3. Therefore,

$$y = \frac{r}{h} x$$

$$\frac{dy}{dx} = \frac{r}{h}$$

$$s = \sqrt{h^2 + r^2}$$

$$S = 2\pi \int_a^b f(x) \sqrt{1 + \left(\frac{dy}{dx}\right)^2}\, dx$$

Fig. 10.3 The geometry
used to compute the surface
area of a right cone

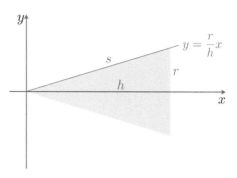

Fig. 10.4 Surface area of a
right cone

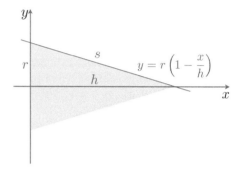

$$= 2\pi \int_0^h \frac{r}{h} x \sqrt{1 + \frac{r^2}{h^2}} \, dx$$

$$= \frac{2\pi r}{h} \int_0^h x \sqrt{\frac{h^2 + r^2}{h^2}} \, dx$$

$$= \frac{2\pi r}{h^2} \int_0^h x \sqrt{h^2 + r^2} \, dx$$

$$= \frac{2\pi r s}{h^2} \int_0^h x \, dx$$

$$= \frac{2\pi r s}{h^2} \cdot \frac{1}{2} x^2 \Big|_0^h$$

$$= \frac{2\pi r s}{h^2} \frac{1}{2} h^2$$

$$= \pi r s$$

which is correct.

Reversing the line's slope to $y = r(1 - x/h)$ as shown in Fig. 10.4 we have

$$y = r \left(1 - \frac{x}{h} \right)$$

Fig. 10.5 The surface of a right cone created by sweeping a line about the x-axis

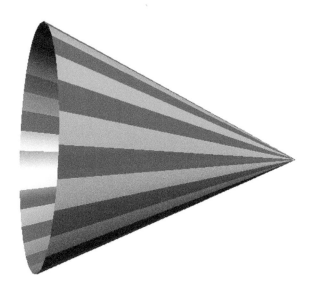

$$\frac{dy}{dx} = -\frac{r}{h}$$

$$s = \sqrt{h^2 + r^2}$$

$$S = 2\pi r \int_0^h \left(1 - \frac{x}{h}\right)\sqrt{1 + \frac{r^2}{h^2}}\, dx$$

$$= \frac{2\pi r}{h}\int_0^h (h - x)\frac{\sqrt{h^2 + r^2}}{h}\, dx$$

$$= \frac{2\pi r s}{h^2} \cdot \left(hx - \tfrac{1}{2}x^2\right)\Big|_0^h$$

$$= \frac{2\pi r s}{h^2}\left(h^2 - \tfrac{1}{2}h^2\right)$$

$$= \frac{2\pi r s}{h^2}\tfrac{1}{2}h^2$$

$$= \pi r s.$$

Figure 10.5 shows a view of the swept conical surface.

10.2.3 Surface Area of a Sphere

The surface area of a sphere is $S = 4\pi r^2$, and is derived as follows.

Figure 10.6 shows a unit semi-circle and Fig. 10.7 shows the surface of revolution when this is swept $360°$ about the x-axis. The equation of a circle is $x^2 + y^2 = r^2$

Fig. 10.6 A unit semi-circle

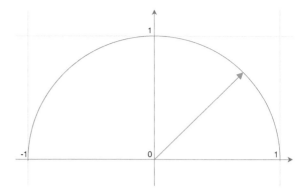

Fig. 10.7 The surface of
revolution formed by
sweeping a semi-circle
through 360°

over the interval $x \in [-r, r]$ therefore,

$$f(x) = y = \sqrt{r^2 - x^2}.$$

To find $f'(x)$, let

$$u = r^2 - x^2$$
$$\frac{du}{dx} = -2x$$
$$y = \sqrt{u}$$
$$\frac{dy}{du} = \tfrac{1}{2}u^{-1/2} = \frac{1}{2\sqrt{u}} = \frac{1}{2\sqrt{r^2 - x^2}}$$

$$\frac{dy}{dx} = \frac{dy}{du} \cdot \frac{du}{dx} = \frac{1}{2\sqrt{r^2 - x^2}}(-2x) = \frac{-x}{\sqrt{r^2 - x^2}}$$

which is substituted in (10.1):

$$S = 2\pi \int_a^b f(x)\sqrt{1 + \left(\frac{dy}{dx}\right)^2}\, dx$$

$$= 2\pi \int_{-r}^r \sqrt{r^2 - x^2}\sqrt{1 + \left(\frac{-x}{\sqrt{r^2 - x^2}}\right)^2}\, dx$$

$$= 2\pi \int_{-r}^r \sqrt{r^2 - x^2}\sqrt{1 + \left(\frac{x^2}{r^2 - x^2}\right)}\, dx$$

$$= 2\pi \int_{-r}^r \sqrt{r^2 - x^2}\frac{r}{\sqrt{r^2 - x^2}}\, dx$$

$$= 2\pi r \int_{-r}^r 1\, dx$$

$$= 2\pi r \cdot x \Big|_{-r}^r$$

$$= 2\pi r 2r$$

$$= 4\pi r^2.$$

10.2.4 Surface Area of a Paraboloid

To compute the surface area of a paraboloid we rotate the parabola function $y = x^2$ about the y-axis, as shown in Fig. 10.8.

$$y = x^2$$

$$x = \sqrt{y}$$

$$\frac{dx}{dy} = \frac{1}{2\sqrt{y}}$$

$$S = 2\pi \int_a^b f(y)\sqrt{1 + \left(\frac{dx}{dy}\right)^2}\, dy$$

$$= 2\pi \int_0^1 \sqrt{y}\sqrt{1 + \left(\frac{1}{2\sqrt{y}}\right)^2}\, dy$$

$$= 2\pi \int_0^1 \sqrt{y}\sqrt{1 + \frac{1}{4y}}\, dy$$

Fig. 10.8 A parabola to be
rotated about the y-axis

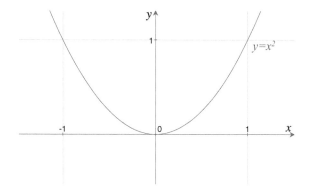

$$= 2\pi \int_0^1 \sqrt{y}\sqrt{\frac{4y+1}{4y}}\, dy$$

$$= 2\pi \int_0^1 \sqrt{y}\frac{\sqrt{4y+1}}{2\sqrt{y}}\, dy$$

$$= \pi \int_0^1 \sqrt{4y+1}\, dy.$$

Let $u = 4y + 1$, therefore, $du/dy = 4$, or $dy = du/4$. The limits for u are 1 and 5.

$$S = \frac{\pi}{4}\int_1^5 \sqrt{u}\, du$$

$$= \frac{\pi}{4}\int_1^5 u^{1/2}\, du$$

$$= \frac{\pi}{4}\cdot \tfrac{2}{3}u^{3/2}\Big|_1^5$$

$$= \frac{\pi}{6}\left(\sqrt{5^3} - \sqrt{1^3}\right)$$

$$= \frac{\pi}{6}\left(\sqrt{125} - 1\right)$$

$$\approx 5.33.$$

Figure 10.9 shows a similar parabolic surface.

10.3   Surface Area Using Parametric Functions

The standard equation to compute surface area is

Fig. 10.9 A parabolic surface

$$S = 2\pi \int_a^b f(x)\sqrt{1 + \left(\frac{dy}{dx}\right)^2}\, dx \qquad (10.4)$$

where the curve represented by $f(x)$ is rotated about the x-axis. In order to convert (10.4) to accept the following parametric equations

$$x = f_x(t)$$
$$y = f_y(t).$$

First, we need to establish equivalent limits $[\alpha, \beta]$ for t, such that

$$a = f_x(\alpha) \quad \text{and} \quad b = f_y(\beta).$$

Second, any point on the curve has corresponding Cartesian and parametric coordinates:

$$x \quad \text{and} \quad f_x(t)$$

$$y = f(x) \quad \text{and} \quad f_y(t).$$

Third, we compute dy/dx from the individual derivatives dx/dt and dy/dt:

$$\frac{dy}{dx} = \frac{dy}{dt} \cdot \frac{dt}{dx}$$

which means that (10.4) can be written as

$$S = 2\pi \int_\alpha^\beta f_y(t)\sqrt{1 + \left(\frac{dy}{dt} \cdot \frac{dt}{dx}\right)^2}\, dx$$

$$= 2\pi \int_\alpha^\beta f_y(t)\sqrt{\frac{(dxdt)^2 + (dydt)^2}{(dxdt)^2}}\, dx$$

$$= 2\pi \int_\alpha^\beta f_y(t) \sqrt{\left(\frac{dx}{dt}\right)^2 + \left(\frac{dy}{dt}\right)^2} \frac{dt}{dx} \, dx$$

$$S = 2\pi \int_\alpha^\beta f_y(t) \sqrt{\left(\frac{dx}{dt}\right)^2 + \left(\frac{dy}{dt}\right)^2} \, dt. \tag{10.5}$$

For example, to create a unit-sphere from the parametric equations for a semi-circle we have

$$x = f_x(t) = -\cos t$$
$$y = f_y(t) = \sin t$$
$$\frac{dx}{dt} = \sin t$$
$$\frac{dy}{dt} = \cos t$$

$$S = 2\pi \int_\alpha^\beta f_y(t) \sqrt{\left(\frac{dx}{dt}\right)^2 + \left(\frac{dy}{dt}\right)^2} \, dt$$

$$= 2\pi \int_0^\pi \sin t \sqrt{\sin^2 t + \cos^2 t} \, dt$$

$$= 2\pi \int_0^\pi \sin t \, dt$$

$$= -2\pi \cdot \cos t \, \Big|_0^\pi$$

$$= 2\pi(1 + 1)$$

$$= 4\pi$$

which is correct.

To rotate about the y-axis (10.5) becomes

$$S = 2\pi \int_\alpha^\beta f_x(t) \sqrt{\left(\frac{dx}{dt}\right)^2 + \left(\frac{dy}{dt}\right)^2} \, dt.$$

10.4   Double Integrals

Up to this point we have only employed single integrals to compute area, but just as it is possible to differentiate a function several times, it is also possible to integrate a function several times. For example, to integrate

$$z = f(x, y) = x^2 y$$

over the interval $x \in [0, 3]$, then we write

$$\int_0^3 f(x, y)\, dx = \int_0^3 x^2 y\, dx$$
$$= \tfrac{1}{3} x^3 y \Big|_0^3$$
$$= 9y.$$

But say we now want to integrate $9y$ over the interval $y \in [0, 2]$, we write

$$\int_0^2 9y\, dy = 9 \int_0^2 y\, dy$$
$$= 9 \cdot \tfrac{1}{2} y^2 \Big|_0^2$$
$$= 18.$$

These two individual steps can be combined in the form of a double integral:

$$\int_0^2 \int_0^3 x^2 y\, dx\, dy$$

where the inner integral is evaluated first, followed by the outer integral:

$$\int_0^2 \int_0^3 x^2 y\, dx\, dy = \int_0^2 \tfrac{1}{3} x^3 \Big|_0^3 y\, dy$$
$$= 9 \int_0^2 y\, dy$$
$$= 9 \cdot \tfrac{1}{2} y^2 \Big|_0^2$$
$$= 18.$$

Note that reversing the integrals has no effect on the result:

$$\int_0^3 \int_0^2 x^2 y\, dy\, dx = \int_0^3 \tfrac{1}{2} y^2 \Big|_0^2 x^2\, dx$$
$$= 2 \int_0^3 x^2\, dx$$
$$= 2 \cdot \tfrac{1}{3} x^3 \Big|_0^3$$
$$= 18.$$

Let's take another example:

$$\int_0^2 \int_1^2 3xy^3 \, dx \, dy = 3 \int_0^2 \tfrac{1}{2}x^2 \Big|_1^2 y^3 \, dy$$

$$= \tfrac{9}{2} \int_0^2 y^3 \, dy$$

$$= \tfrac{9}{2} \cdot \tfrac{1}{4}y^4 \Big|_0^2$$

$$= 18.$$

10.5   Jacobians

In spite of a relatively short life, the German mathematician Carl Gustav Jacob Jacobi (1804–1851) made a significant contribution to mathematics in the areas of elliptic functions, number theory, differential equations and in particular, the Jacobian matrix and determinant.

The *Jacobian matrix* is used in equations of differentials when changing variables, and its determinant, the *Jacobian determinant*, provides a scaling factor in multiple integrals when changing the independent variable. I will provide three applications of the determinant, showing its use in one, two and three dimensions.

10.5.1   1D Jacobian

In order to integrate some integrals, we often have to substitute a new variable. For example, to integrate

$$\int_1^4 \sqrt{2x + 1} \, dx$$

it is convenient to substitute $u = 2x + 1$, where $du/dx = 2$ or $dx/du = 1/2$, calculate new limits for u: i.e. 3 and 9, and integrate with respect to u:

$$\int_1^4 \sqrt{2x + 1} \, dx = \int_3^9 \sqrt{u} \, \frac{dx}{du} du$$

$$= \tfrac{1}{2} \int_3^9 \sqrt{u} \, du$$

$$= \tfrac{1}{2} \int_3^9 u^{1/2} \, du$$

$$= \tfrac{1}{2} \cdot \tfrac{2}{3}u^{3/2} \Big|_3^9$$

$$= \tfrac{1}{3}\left(9^{3/2} - 3^{3/2}\right)$$
$$\approx \tfrac{1}{3}(27 - 5.2)$$
$$\approx 7.3.$$

The factor $1/2$ is introduced because x changes half as fast as u. This scaling factor is known as a *Jacobian*, and is the derivative dx/du. We can also write it as $\partial x/\partial u$, even though there is only one variable, as the partial notation keeps the Jacobians consistent as we increase the number of dimensions. Furthermore, we are only interested in the magnitude of the Jacobian, not its sign.

The scaling factor could also be another function. For example, to integrate

$$\int_0^2 \frac{x}{(x^2 + 2)^2}\, dx$$

it is convenient to substitute $u = x^2 + 2$, where $du/dx = 2x$ or $dx/du = 1/2x$, calculate new limits for u: i.e. 2 and 6, and integrate with respect to u:

$$\int_0^2 \frac{x}{(x^2 + 2)^2}\, dx = \int_2^6 \frac{x}{u^2} \frac{dx}{du} du$$
$$= \frac{1}{2x} \int_2^6 \frac{x}{u^2}\, du$$
$$= \tfrac{1}{2} \int_2^6 \frac{1}{u^2}\, du$$
$$= \tfrac{1}{2} \int_2^6 u^{-2}\, du$$
$$= \tfrac{1}{2} \cdot \left. \frac{-1}{u} \right|_2^6$$
$$= \tfrac{1}{2}\left(-\tfrac{1}{6} + \tfrac{1}{2}\right)$$
$$= \tfrac{1}{6}.$$

In this case, the scaling factor is $1/2x$, which is the corresponding Jacobian, however, this time its value is a function of x.

10.5.2   2D Jacobian

When defining double integrals using Cartesian coordinates, one normally ends up with something like

Fig. 10.10 The rectangle $C_1 C_2 C_3 C_4$ in Cartesian space

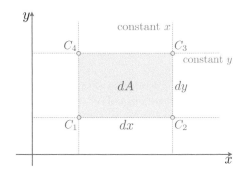

Fig. 10.11 The rectangle $P_1 P_2 P_3 P_4$ in parametric space

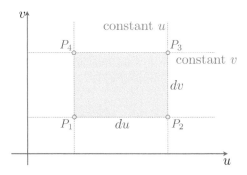

$$\int_a^b f(x, y) \, dx \, dy$$

where $dx \, dy$ is regarded as the area of an infinitesimally small rectangle, and is often represented by dA. But if we move from Cartesian coordinates to polar coordinates and work with functions of the form $g(\rho, \theta)$, there is a temptation to substitute $g(\rho, \theta)$ for $f(x, y)$ and $(d\rho \, d\theta)$ for $(dx \, dy)$, which is incorrect. The reason why, is that the differential area of a rectangular region in Cartesian coordinates does **not** equal the differential area of a corresponding region in polar coordinates. The Jacobian determinant provides us with the adjustment necessary to carry out this substitution, which in this case is ρ, and $(dx \, dy)$ is replaced by $(\rho \, d\rho \, d\theta)$. I will describe a general solution to this problem, which is found on various internet websites, but in particular http://mathforum.org/dr.math/.

Figure 10.10 shows an infinitesimally small rectangle defined by the points $C_1 C_2 C_3 C_4$ in Cartesian coordinates. The vertical broken lines identify lines of constant x, and the horizontal broken lines identify lines of constant y. The rectangle's width and height are dx and dy, respectively, which makes $dA = dx \, dy$. Similarly, Fig. 10.11 shows an infinitesimally small rectangle defined by the points $P_1 P_2 P_3 P_4$ in another coordinate system. The vertical broken lines identify lines of constant u, and the horizontal broken lines identify lines of constant v. The rectangle's width and height are du and dv, respectively.

Fig. 10.12 The parametric
points $P_1 P_2 P_3 P_4$ in
Cartesian space

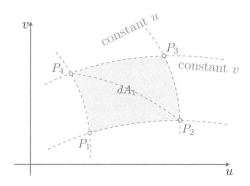

We now create two single-valued functions mapping parametric coordinates (u, v)
into Cartesian coordinates (x, y):

$$x = f(u, v) \quad \text{and} \quad y = g(u, v)$$

where for every (x, y) there is a unique (u, v). There are also two single-valued
functions mapping Cartesian coordinates (x, y) into parametric coordinates (u, v):

$$u = F(x, y) \quad \text{and} \quad v = G(x, y).$$

For example, given

$$u = x^2 + y^2 \quad \text{and} \quad v = x^2 - y^2$$

then

$$x = \sqrt{\frac{u + v}{2}} \quad \text{and} \quad y = \sqrt{\frac{u - v}{2}}.$$

Next, we take the points in uv-space and map them into their corresponding Cartesian
points as shown in Fig. 10.12. The resulting shape depends entirely upon the nature
of the mapping functions $f(u, v)$ and $g(u, v)$; however, we anticipate that they are
curved in some way and bounded by contours of constant u and v.

If the area of this differential region equals the Cartesian rectangle $dx \, dy$, then
$dx \, dy$ can be replaced by $du \, dv$. If not, we must compensate for any stretching or
contraction. The problem therefore, is to compute the area of this curvilinear rectangle
$P_1 P_2 P_3 P_4$ in Fig. 10.12 and compare it to the area of the rectangle $C_1 C_2 C_3 C_4$ in
Fig. 10.10. This is resolved by assuming that when this rectangle is infinitesimally
small, curves can be approximated by lines, and the area of the triangle $P_1 P_2 P_4$ is
half the area of the required region. The area of the triangle is easily computed using
the determinant

$$\frac{1}{2} \begin{vmatrix} 1 & 1 & 1 \\ x_1 & x_2 & x_4 \\ y_1 & y_2 & y_4 \end{vmatrix}$$

where (x_1, y_1), (x_2, y_2) and (x_4, y_4) are the triangle's vertices taken in anticlockwise sequence. Reversing the sequence, reverses the sign, which is why the absolute value is added at the end of the proof. However, if we assume that the area of the curvilinear region is twice the area of the triangle, then

$$\text{Area of } (P_1 P_2 P_3 P_4) = dA_1 = \begin{vmatrix} 1 & 1 & 1 \\ x_1 & x_2 & x_4 \\ y_1 & y_2 & y_4 \end{vmatrix}. \tag{10.6}$$

The next stage is to derive a function relating the differentials dx and dy with du and dv, so that the triangle's coordinates can be determined. These functions are simply the total differentials for f and g:

$$x = f(u, v)$$
$$y = g(u, v)$$
$$dx = \frac{\partial x}{\partial u} du + \frac{\partial x}{\partial v} dv$$
$$dy = \frac{\partial y}{\partial u} du + \frac{\partial y}{\partial v} dv.$$

As with many mathematical solutions we can save ourselves a lot of work by making a simple assumption, which in this case is that the coordinates of P_1 are (x_1, y_1), and the coordinates of P_2 and P_4 are of the form $(x_1 + dx, y_1 + dy)$.

Starting with P_2 with coordinates (x_2, y_2), then

$$x_2 = x_1 + dx$$
$$y_2 = y_1 + dy$$
$$x_2 = x_1 + \frac{\partial x}{\partial u} du + \frac{\partial x}{\partial v} dv$$
$$y_2 = y_1 + \frac{\partial y}{\partial u} du + \frac{\partial y}{\partial v} dv$$

but as P_1 and P_2 lie on a contour where v is constant, $dv = 0$, which means that

$$x_2 = x_1 + \frac{\partial x}{\partial u} du$$
$$y_2 = y_1 + \frac{\partial y}{\partial u} du.$$

Next, P_4 with coordinates (x_4, y_4), then

$$x_4 = x_1 + dx$$
$$y_4 = y_1 + dy$$

$$x_4 = x_1 + \frac{\partial x}{\partial u}du + \frac{\partial x}{\partial v}dv$$

$$y_4 = y_1 + \frac{\partial y}{\partial u}du + \frac{\partial y}{\partial v}dv$$

but as P_1 and P_4 lie on a contour where u is constant, $du = 0$, which means that

$$x_4 = x_1 + \frac{\partial x}{\partial v}dv$$

$$y_4 = y_1 + \frac{\partial y}{\partial v}dv.$$

We now plug the coordinates for P_1, P_2 and P_4 into (10.6):

$$dA_1 = \begin{vmatrix} 1 & 1 & 1 \\ x_1 & x_1 + \dfrac{\partial x}{\partial u}du & x_1 + \dfrac{\partial x}{\partial v}dv \\ y_1 & y_1 + \dfrac{\partial y}{\partial u}du & y_1 + \dfrac{\partial y}{\partial v}dv \end{vmatrix}.$$

Rather than expand the determinant, let's simplify it by subtracting column 1 from columns 2 and 3:

$$dA_1 = \begin{vmatrix} 1 & 0 & 0 \\ x_1 & \dfrac{\partial x}{\partial u}du & \dfrac{\partial x}{\partial v}dv \\ y_1 & \dfrac{\partial y}{\partial u}du & \dfrac{\partial y}{\partial v}dv \end{vmatrix}$$

which becomes

$$dA_1 = \begin{vmatrix} \dfrac{\partial x}{\partial u}du & \dfrac{\partial x}{\partial v}dv \\ \dfrac{\partial y}{\partial u}du & \dfrac{\partial y}{\partial v}dv \end{vmatrix}.$$

The determinant now contains the common term $du\,dv$, which is taken outside:

$$dA_1 = \begin{vmatrix} \dfrac{\partial x}{\partial u} & \dfrac{\partial x}{\partial v} \\ \dfrac{\partial y}{\partial u} & \dfrac{\partial y}{\partial v} \end{vmatrix} du\,dv.$$

Finally, we write this as

$$dA_1 = \frac{\partial(x, y)}{\partial(u, v)}du\,dv = |J|\,du\,dv$$

where J is the Jacobian determinant

$$J = \begin{vmatrix} \dfrac{\partial x}{\partial u} & \dfrac{\partial x}{\partial v} \\ \dfrac{\partial y}{\partial u} & \dfrac{\partial y}{\partial v} \end{vmatrix}.$$

Therefore, for the region R, we can write

$$\iint_{R(x,y)} F(x, y) \, dx \, dy = \iint_{R(u,v)} F\left(f(u, v), g(u, v)\right) \, |J| \, du \, dv$$

Let's evaluate J for converting Cartesian to polar coordinates, where

$$x = \rho \cos \theta$$
$$y = \rho \sin \theta$$

therefore,

$$\frac{\partial x}{\partial \rho} = \cos \theta, \quad \frac{\partial x}{\partial \theta} = -\rho \sin \theta, \quad \frac{\partial y}{\partial \rho} = \sin \theta, \quad \frac{\partial y}{\partial \theta} = \rho \cos \theta,$$

$$J = \begin{vmatrix} \cos \theta & -\rho \sin \theta \\ \sin \theta & \rho \cos \theta \end{vmatrix}$$
$$= \rho \cos^2 \theta + \rho \sin^2 \theta$$
$$= \rho$$

therefore, $dx \, dy$ is replaced by $\rho \, d\rho \, d\theta$.

10.5.3 3D Jacobian

The Jacobian determinant generalises to higher dimensions, and in three dimensions becomes

$$J = \begin{vmatrix} \dfrac{\partial x}{\partial u} & \dfrac{\partial x}{\partial v} & \dfrac{\partial x}{\partial w} \\ \dfrac{\partial y}{\partial u} & \dfrac{\partial y}{\partial v} & \dfrac{\partial y}{\partial w} \\ \dfrac{\partial z}{\partial u} & \dfrac{\partial z}{\partial v} & \dfrac{\partial z}{\partial w} \end{vmatrix} \tag{10.7}$$

and is used in with triple integrals for calculating volumes. For example, in the next chapter I will show how a triple integral using spherical coordinates is converted into Cartesian coordinates using the appropriate Jacobian. For the moment, let's evaluate the Jacobian determinant. Figure 10.13 shows the convention used for converting the point (x, y, z) into spherical polar coordinates (ρ, ϕ, θ). From Fig. 10.13 we see that

Fig. 10.13 Spherical polar
coordinates

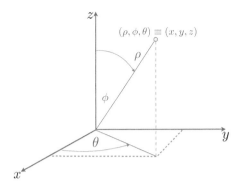

$$x = \rho \sin \phi \cdot \cos \theta$$
$$y = \rho \sin \phi \cdot \sin \theta$$
$$z = \rho \cos \phi$$

the partial derivatives are

$$\frac{\partial x}{\partial \rho} = \sin \phi \cdot \cos \theta, \quad \frac{\partial x}{\partial \phi} = \rho \cos \phi \cdot \cos \theta, \quad \frac{\partial x}{\partial \theta} = -\rho \sin \phi \cdot \sin \theta,$$

$$\frac{\partial y}{\partial \rho} = \sin \phi \cdot \sin \theta, \quad \frac{\partial y}{\partial \phi} = \rho \cos \phi \cdot \sin \theta, \quad \frac{\partial y}{\partial \theta} = \rho \sin \phi \cdot \cos \theta,$$

$$\frac{\partial z}{\partial \rho} = \cos \phi, \quad \frac{\partial z}{\partial \phi} = -\rho \sin \phi, \quad \frac{\partial z}{\partial \theta} = 0.$$

Substituting these partials in (10.7):

$$J = \begin{vmatrix} \sin \phi \cdot \cos \theta & \rho \cos \phi \cdot \cos \theta & -\rho \sin \phi \cdot \sin \theta \\ \sin \phi \cdot \sin \theta & \rho \cos \phi \cdot \sin \theta & \rho \sin \phi \cdot \cos \theta \\ \cos \phi & -\rho \sin \phi & 0 \end{vmatrix}$$

which expands to

$$\det = \rho^2 \cos^2 \phi \cdot \cos^2 \theta \cdot \sin \phi + \rho^2 \sin^3 \phi \cdot \sin^2 \theta + \rho^2 \sin^3 \phi \cdot \cos^2 \theta + \rho^2 \sin \phi \cdot \sin^2 \theta \cdot \cos^2 \phi$$
$$= \left(\rho^2 \sin^3 \phi + \rho^2 \sin \phi \cdot \cos^2 \phi \right)(\sin^2 \theta + \cos^2 \theta)$$
$$= \rho^2 \sin \phi \left(\sin^2 \phi + \cos^2 \phi \right)$$
$$= \rho^2 \sin \phi.$$

Normally, we take the absolute value of the Jacobian determinant, but in this case, $\phi \in [0, \pi]$, and $\rho^2 \sin \phi$ is always positive. Thus $\rho^2 \sin \phi \, d\phi \, d\theta$ replaces $dx \, dy \, dz$ in the appropriate integral.

When using cylindrical coordinates, where

$$x = \rho \cos \phi, \quad y = \rho \sin \phi, \quad z = z,$$

the Jacobian is ρ:

$$
\begin{aligned}
J &= \begin{vmatrix}
\dfrac{\partial x}{\partial u} & \dfrac{\partial x}{\partial v} & \dfrac{\partial x}{\partial w} \\[2mm]
\dfrac{\partial y}{\partial u} & \dfrac{\partial y}{\partial v} & \dfrac{\partial y}{\partial w} \\[2mm]
\dfrac{\partial z}{\partial u} & \dfrac{\partial z}{\partial v} & \dfrac{\partial z}{\partial w}
\end{vmatrix} \\[3mm]
&= \begin{vmatrix}
\cos \phi & -\rho \sin \phi & 0 \\
\sin \phi & \rho \cos \phi & 0 \\
0 & 0 & 1
\end{vmatrix} \\[2mm]
&= \rho \cos^2 \phi + \rho \sin^2 \phi \\[1mm]
&= \rho.
\end{aligned}
$$

Thus the first three Jacobians are

$$
J_1 = \frac{\partial x}{\partial u}, \quad
J_2 = \begin{vmatrix}
\dfrac{\partial x}{\partial u} & \dfrac{\partial x}{\partial v} \\[2mm]
\dfrac{\partial y}{\partial u} & \dfrac{\partial y}{\partial v}
\end{vmatrix}, \quad
J_3 = \begin{vmatrix}
\dfrac{\partial x}{\partial u} & \dfrac{\partial x}{\partial v} & \dfrac{\partial x}{\partial w} \\[2mm]
\dfrac{\partial y}{\partial u} & \dfrac{\partial y}{\partial v} & \dfrac{\partial y}{\partial w} \\[2mm]
\dfrac{\partial z}{\partial u} & \dfrac{\partial z}{\partial v} & \dfrac{\partial z}{\partial w}
\end{vmatrix}
$$

which are often compressed to

$$
J_1 = \frac{\partial x}{\partial u}, \quad
J_2 = \frac{\partial(x, y)}{\partial(u, v)}, \quad
J_3 = \frac{\partial(x, y, z)}{\partial(u, v, w)}.
$$

10.6   Double Integrals for Calculating Area

I will now illustrate how double integrals are used for calculating area, and in the next chapter, show how they are also used for calculating volume. To begin, look what happens when we integrate $f(x, y) = 1$ over the interval $x \in [a, b]$, and $y \in [c, d]$:

$$
\begin{aligned}
\int_c^d \int_a^b f(x, y)\, dx\, dy &= \int_c^d \int_a^b 1 \, dx\, dy \\[2mm]
&= \int_c^d x \Big|_a^b \, dy
\end{aligned}
$$

Fig. 10.14 The projection of $z = f(x, y)$ on the xy-plane

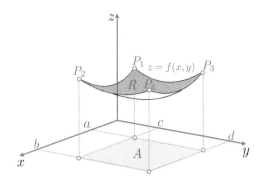

$$= \int_c^d (b - a)\, dy$$

$$= (b - a) \int_c^d 1\, dy$$

$$= (b - a) \cdot y \Big|_c^d$$

$$= (b - a)(d - c).$$

The result is the product of the x- and y-intervals, which is the region A formed by a 3D surface projected onto the xy-plane, as shown in Fig. 10.14. The actual area of the surface created by $z = f(x, y)$ bounded by the points P_1, P_2, P_3 and P_4 is given by

$$R = \int_c^d \int_a^b \sqrt{1 + \left(\frac{\partial z}{\partial x}\right)^2 + \left(\frac{\partial z}{\partial y}\right)^2}\, dx\, dy. \qquad (10.8)$$

Let's show how (10.8) is used to compute area. The first example is simple and is shown in Fig. 10.15, where $z = f(x, y) = y$. The intervals are $x \in [0, 2]$ and $y \in [0, 1]$. By inspection, the area equals $2\sqrt{2}$. Calculating the partial derivatives, we have

$$\frac{\partial z}{\partial x} = 0, \quad \text{and} \quad \frac{\partial z}{\partial y} = 1$$

therefore, (10.8) becomes

$$R = \int_0^1 \int_0^2 \sqrt{1 + 0^2 + 1^2}\, dx\, dy$$

$$= \sqrt{2} \int_0^1 \int_0^2 1\, dx\, dy$$

Fig. 10.15 Part of the
surface $z = y$

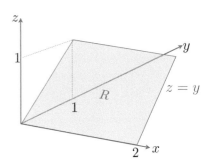

Fig. 10.16 Part of the
surface $z = 4x + 2y$

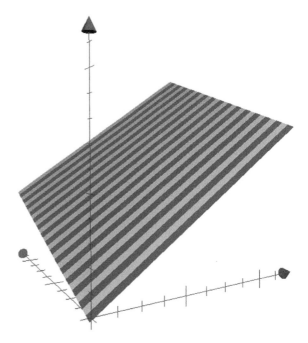

$$= \sqrt{2} \int_0^1 x \Big|_0^2 \, dy$$

$$= 2\sqrt{2} \int_0^1 1 \, dy$$

$$= 2\sqrt{2} \cdot y \Big|_0^1$$

$$= 2\sqrt{2}.$$

The second example is shown in Fig. 10.16, where $z = f(x, y) = 4x + 2y$. The intervals are $x \in [0, 1]$ and $y \in [0, 1]$. Calculating the partial derivatives, we have

$$\frac{\partial z}{\partial x} = 4, \quad \text{and} \quad \frac{\partial z}{\partial y} = 2,$$

therefore, (10.8) becomes

$$R = \int_0^1 \int_0^1 \sqrt{1 + 4^2 + 2^2} \, dx \, dy$$

$$= \sqrt{21} \int_0^1 \int_0^1 1 \, dx \, dy$$

$$= \sqrt{21} \int_0^1 x \Big|_0^1 \, dy$$

$$= \sqrt{21} \int_0^1 1 \, dy$$

$$= \sqrt{21} \cdot y \Big|_0^1$$

$$= \sqrt{21}.$$

We can also calculate the area of the surface $z = 4x + 2y$ contained within a specific region on the xy-plane as follows. For example, say the region is defined by

$$x^2 + y^2 = 1$$

as shown in Fig. 10.17, we calculate the area as follows.

To begin, we use polar coordinates instead of Cartesian coordinates, incorporating the vital Jacobian, and rewrite (10.8) as

$$R = \int_0^{\pi/2} \int_0^1 \sqrt{1 + \left(\frac{\partial z}{\partial x}\right)^2 + \left(\frac{\partial z}{\partial y}\right)^2} \, \rho \, d\rho \, d\theta. \tag{10.9}$$

The inner integral integrates over the interval $\rho \in [0, 1]$, and the outer integral integrates over the interval $\theta \in [0, \pi/2]$. Using the same equations, we have

$$R = \int_0^{\pi/2} \int_0^1 \sqrt{1 + \left(\frac{\partial z}{\partial x}\right)^2 + \left(\frac{\partial z}{\partial y}\right)^2} \, \rho \, d\rho \, d\theta$$

$$= \int_0^{\pi/2} \int_0^1 \sqrt{1 + 4^2 + 2^2} \, \rho \, d\rho \, d\theta$$

$$= \sqrt{21} \int_0^{\pi/2} \int_0^1 \rho \, d\rho \, d\theta$$

$$= \sqrt{21} \int_0^{\pi/2} \rho \Big|_0^1 \, d\theta$$

Fig. 10.17 The graph of
$z = 4x + 2y$ intersecting the
cylinder defined by
$x^2 + y^2 = 1$ on the xy-plane

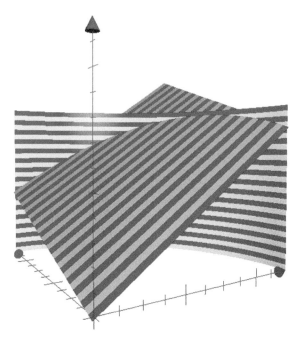

$$= \sqrt{21} \int_0^{\pi/2} 1 \, d\theta$$

$$= \sqrt{21} \cdot \theta \Big|_0^{\pi/2}$$

$$= \frac{\sqrt{21}\pi}{2}$$

$$\approx 7.2.$$

For a third example, Fig. 10.18 shows part of a cone $z = 4\sqrt{x^2 + y^2}$ intersecting a
cylinder defined by $x^2 + y^2 = 1$ on the xy-plane. Let's calculate the area of the cone
contained within the cylindrical region over $\rho \in [0, 1]$, and $\theta \in [0, \pi/2]$.

The partial derivatives are

$$\frac{\partial z}{\partial x} = \frac{4x}{\sqrt{x^2 + y^2}}, \quad \text{and} \quad \frac{\partial z}{\partial y} = \frac{4y}{\sqrt{x^2 + y^2}},$$

therefore, using (10.9) we have

$$R = \int_0^{\pi/2} \int_0^1 \sqrt{1 + \left(\frac{\partial z}{\partial x}\right)^2 + \left(\frac{\partial z}{\partial y}\right)^2} \, \rho \, d\rho \, d\theta$$

Fig. 10.18 The graph of
$z = 4\sqrt{x^2 + y^2}$ intersecting
the cylinder defined by
$x^2 + y^2 = 1$ on the xy-plane

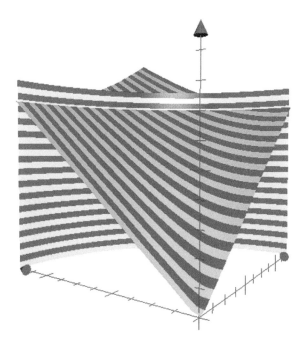

$$= \int_0^{\pi/2} \int_0^1 \sqrt{1 + \left(\frac{4x}{\sqrt{x^2 + y^2}}\right)^2 + \left(\frac{4y}{\sqrt{x^2 + y^2}}\right)^2} \, \rho \, d\rho \, d\theta$$

$$= \int_0^{\pi/2} \int_0^1 \sqrt{1 + \frac{16x^2}{x^2 + y^2} + \frac{16y^2}{x^2 + y^2}} \, \rho \, d\rho \, d\theta$$

$$= \sqrt{17} \int_0^{\pi/2} \int_0^1 \rho \, d\rho \, d\theta$$

$$= \sqrt{17} \int_0^{\pi/2} \rho \, \Big|_0^1 \, d\theta$$

$$= \sqrt{17} \int_0^{\pi/2} 1 \, d\theta$$

$$= \sqrt{17} \cdot \theta \, \Big|_0^{\pi/2}$$

$$= \frac{\sqrt{17}\pi}{2}$$

$$\approx 6.48.$$

The above examples have been carefully chosen so that the radical within the integrand reduces to some numerical value. Unfortunately, this is not always the case, and integration has to involve software or numerical methods.

10.7   Summary

In this chapter we have derived formulae to compute the surface area of contours rotated about the x- and y-axis. The important formulae are repeated below.

10.7.1   Summary of Formulae

Rotate About the x-axis

$$S = 2\pi \int_a^b f(x)\sqrt{1 + \left(\frac{dy}{dx}\right)^2}\, dx.$$

Rotate About the y-axis

$$S = 2\pi \int_a^b f(y)\sqrt{1 + \left(\frac{dx}{dy}\right)^2}\, dy.$$

If the function is described parametrically with $x = f_x(t)$ and $y = f_y(t)$ where $t \in [\alpha, \beta]$, then:

Rotate About the x-axis

$$S = 2\pi \int_\alpha^\beta f_y(t)\sqrt{\left(\frac{dx}{dt}\right)^2 + \left(\frac{dy}{dt}\right)^2}\, dt.$$

Rotate About the y-axis

$$S = 2\pi \int_\alpha^\beta f_x(t)\sqrt{\left(\frac{dx}{dt}\right)^2 + \left(\frac{dy}{dt}\right)^2}\, dt.$$

Double integrals for calculating the area of surfaces described by functions of the form $z = f(x, y)$, then:

Cartesian Coordinates

$$R = \int\int_R \sqrt{1 + \left(\frac{\partial z}{\partial x}\right)^2 + \left(\frac{\partial z}{\partial y}\right)^2}\, dx\, dy$$

Spherical Polar Coordinates

$$R = \int\int_R \sqrt{1 + \left(\frac{\partial z}{\partial x}\right)^2 + \left(\frac{\partial z}{\partial y}\right)^2}\, \rho\, d\rho\, d\theta.$$

The First Three Jacobian Determinants

$$J_1 = \frac{\partial x}{\partial u}, \quad J_2 = \begin{vmatrix} \dfrac{\partial x}{\partial u} & \dfrac{\partial x}{\partial v} \\[2mm] \dfrac{\partial y}{\partial u} & \dfrac{\partial y}{\partial v} \end{vmatrix}, \quad J_3 = \begin{vmatrix} \dfrac{\partial x}{\partial u} & \dfrac{\partial x}{\partial v} & \dfrac{\partial x}{\partial w} \\[2mm] \dfrac{\partial y}{\partial u} & \dfrac{\partial y}{\partial v} & \dfrac{\partial y}{\partial w} \\[2mm] \dfrac{\partial z}{\partial u} & \dfrac{\partial z}{\partial v} & \dfrac{\partial z}{\partial w} \end{vmatrix}$$

which are often written as

$$J_1 = \frac{\partial x}{\partial u}, \quad J_2 = \frac{\partial(x, y)}{\partial(u, v)}, \quad J_3 = \frac{\partial(x, y, z)}{\partial(u, v, w)}.$$

Chapter 11
Volume

11.1 Introduction

In this chapter I introduce four techniques for calculating the volume of various geometric objects. Two techniques are associated with solids of revolution, where an object is cut into flat slices or concentric cylindrical shells and summed over the object's extent using a single integral. The third technique employs two integrals where the first computes the area of a slice through a volume, and the second sums these areas over the object's extent. The fourth technique employs three integrals to sum the volume of an object. We start with the slicing technique.

11.2 Solid of Revolution: Disks

In Chap. 10 we saw that the area of a swept surface is calculated using

$$S = 2\pi \int_a^b f(x)\sqrt{1 + \left(\frac{dy}{dx}\right)^2}\, dx.$$

Now let's show that the contained volume is given by

$$V = \pi \int_a^b (f(x))^2 \, dx.$$

Figure 11.1 shows a contour described by $y = f(x)$ rotated about the x-axis creating a solid of revolution. If we imagine this object cut into a series of thin slices, then the entire volume is the sum of the volumes of the individual slices. However, if we cut a real solid of revolution into a collection of slices, it is highly likely that each slice forms a right conical frustum, where the diameter of one side differs slightly from the other side. Therefore, our numerical strategy assumes that

© Springer Nature Switzerland AG 2019
J. Vince, *Calculus for Computer Graphics*,
https://doi.org/10.1007/978-3-030-11376-6_11

Fig. 11.1 Dividing a volume
of revolution into small disks

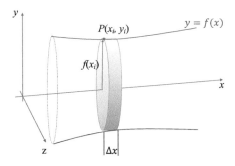

the slices are infinitesimally thin, and are thin disks with a volume equal to $\pi r^2 \Delta x$.
Figure 11.1 shows a point $P(x_i, y_i)$ on the contour touching a disk with radius $f(x_i)$
and thickness Δx. Therefore, the volume of the disk is

$$V_i = \pi \left(f(x_i)\right)^2 \Delta x.$$

Dividing the contour into n such disks, and letting n tend towards infinity, the entire
volume is given by

$$V = \lim_{n \to \infty} \sum_{i=1}^{n} \pi \left(f(x_i)\right)^2 \Delta x$$

which in integral form is

$$V = \pi \int_{a}^{b} \left(f(x)\right)^2 \, dx. \tag{11.1}$$

Let's apply (11.1) to the same objects used for computing the surface area of surfaces
of revolution.

11.2.1   Volume of a Cylinder

The geometry required to compute the volume of a cylinder is shown in Fig. 11.2,
where $y = r$ (the radius) and h is the height. Therefore, using (11.1) we have

$$V = \pi \int_{a}^{b} \left(f(x)\right)^2 \, dx$$

$$= \pi \int_{0}^{h} r^2 \, dx$$

$$= \pi r^2 \int_{0}^{h} 1 \, dx$$

$$= \pi r^2 \cdot x \Big|_{0}^{h}$$

$$= \pi r^2 h.$$

Fig. 11.2 Computing the
volume of a cylinder

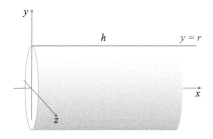

11.2.2   Volume of a Right Cone

The geometry required to compute the volume of a right cone is shown in Fig. 11.3,
where $y = rx/h$. Therefore, using (11.1) we have

$$V = \pi \int_a^b (f(x))^2 \, dx$$

$$= \pi \int_0^h \frac{r^2}{h^2} x^2 \, dx$$

$$= \frac{\pi r^2}{h^2} \int_0^h x^2 \, dx$$

$$= \frac{\pi r^2}{h^2} \cdot \frac{1}{3} x^3 \Big|_0^h$$

$$= \frac{\pi r^2}{h^2} \frac{1}{3} h^3$$

$$= \frac{1}{3} \pi r^2 h.$$

Reversing the orientation of the cone as shown in Fig. 11.4, such that $y = r(1 - x/h)$ we have

$$V = \pi \int_a^b (f(x))^2 \, dx$$

$$= \pi \int_0^h r^2 \left(1 - \frac{x}{h}\right)^2 \, dx$$

$$= \pi r^2 \int_0^h \left(1 - \frac{x}{h}\right)^2 \, dx$$

$$= \pi r^2 \int_0^h \left(1 - \frac{2x}{h} + \frac{x^2}{h^2}\right) \, dx$$

$$= \pi r^2 \left[x - \frac{x^2}{h} + \frac{x^3}{3h^2}\right]_0^h$$

$$= \pi r^2 \left(h - h + \frac{1}{3}h\right)$$

$$= \frac{1}{3} \pi r^2 h.$$

Fig. 11.3 Computing the
volume of a right cone

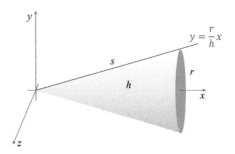

We could have also integrated this as follows:

$$V = \pi \int_a^b (f(x))^2 \; dx$$

$$= \pi \int_0^h r^2 \left(1 - \frac{x}{h}\right)^2 \; dx$$

$$= \pi r^2 \int_0^h \left(1 - \frac{x}{h}\right)^2 \; dx.$$

Substituting

$$u = 1 - \frac{x}{h}$$

where $du/dx = -1/h$, or $dx = -h \, du$, and calculating new limits for u: $[1, 0]$,
we have

$$V = \pi r^2 \int_1^0 u^2(-h) \; du$$

$$= \pi r^2 h \int_0^1 u^2 \; du$$

$$= \pi r^2 h \cdot \frac{1}{3}u^3 \Big|_0^1$$

$$= \tfrac{1}{3}\pi r^2 h.$$

11.2.3   Volume of a Right Conical Frustum

Figure 11.5 shows the geometry to compute the volume of a right conical frustum,
but this time the contour is rotated about the y-axis. The integral to achieve this is

$$V = \pi \int_a^b (f(y))^2 \; dy$$

Fig. 11.4 Reversing the
orientation of a right cone

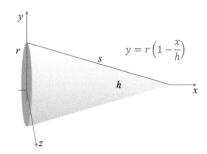

Fig. 11.5 Computing the
volume of a right conical
frustum

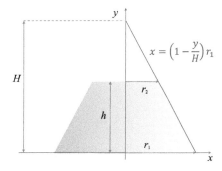

and the contour to be rotated about the y-axis is

$$x = \left(1 - \frac{y}{H}\right) r_1$$

with the integral for the volume:

$$V = \pi r_1^2 \int_0^h \left(1 - \frac{y}{H}\right)^2 dy.$$

However, in reality, we will not know the value of H, but we would know the values
of r_1 and r_2. Therefore, with a little manipulation, the contour can be written as

$$x = \frac{hr_1 + y(r_2 - r_1)}{h}$$

which confirms that when $y = 0$, $x = r_1$, and when $y = h$, $x = r_2$. Therefore, the
volume can be written in terms of r_1, r_2 and h as:

$$V = \frac{\pi}{h^2} \int_0^h (hr_1 + y(r_2 - r_1))^2 \, dy$$

$$= \frac{\pi}{h^2} \int_0^h h^2 r_1^2 + 2hr_1 y(r_2 - r_1) + y^2 (r_2 - r_1)^2 \, dy$$

Fig. 11.6 A semi-circle
used to form a sphere

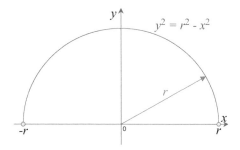

$$= \frac{\pi}{h^2} \left[h^2 r_1^2 y + hr_1 y^2 (r_2 - r_1) + \tfrac{1}{3} y^3 \left(r_2^2 - 2r_1 r_2 + r_1^2 \right) \right]_0^h$$

$$= \frac{\pi}{h^2} \left(h^3 r_1^2 + h^3 r_1 (r_2 - r_1) + \tfrac{1}{3} h^3 \left(r_2^2 - 2r_1 r_2 + r_1^2 \right) \right)$$

$$= \frac{\pi h}{3} \left(3r_1^2 + 3r_1 r_2 - 3r_1^2 + r_2^2 - 2r_1 r_2 + r_1^2 \right)$$

$$= \frac{\pi h}{3} \left(r_1^2 + r_2^2 + r_1 r_2 \right).$$

For example, when $r_1 = 2$ cm, $r_2 = 4$ cm and $h = 3$ cm, then

$$V = \frac{3\pi}{3} \left(2^2 + 4^2 + 8 \right) = 28\pi \text{ cm}^3.$$

11.2.4   Volume of a Sphere

A sphere is easily created by rotating a semi-circle about the x- or y-axis, as shown
in Fig. 11.6, where the equation of the contour is given by

$$y^2 = r^2 - x^2.$$

Using (11.1), the volume is

$$V = \pi \int_{-r}^{r} y^2 \, dx$$

$$= \pi \int_{-r}^{r} \left(r^2 - x^2 \right) dx$$

$$= \pi \left[r^2 x - \tfrac{1}{3} x^3 \right]_{-r}^{r}$$

$$= \pi \left(r^3 - \tfrac{1}{3} r^3 + r^3 - \tfrac{1}{3} r^3 \right)$$

$$= \tfrac{4}{3} \pi r^3.$$

Fig. 11.7 Part of an ellipse
used to form an ellipsoid

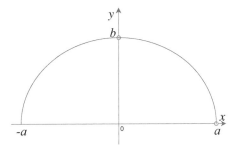

11.2.5   Volume of an Ellipsoid

Figure 11.7 shows part of an ellipse, which when rotated about the x-axis creates a
3D ellipsoid. Using (11.1) with the equation for an ellipse:

$$\left(\frac{x}{a}\right)^2 + \left(\frac{y}{b}\right)^2 = 1$$

we have

$$y^2 = \frac{b^2}{a^2}\left(a^2 - x^2\right)$$

where the ellipsoid's volume is given by

$$
\begin{aligned}
V &= \pi \int_{-a}^{a} y^2 \, dx \\
&= \pi \frac{b^2}{a^2} \int_{-a}^{a} \left(a^2 - x^2\right) \, dx \\
&= \pi \frac{b^2}{a^2} \left[a^2 x - \tfrac{1}{3}x^3\right]_{-a}^{a} \\
&= \pi \frac{b^2}{a^2} \left(a^3 - \tfrac{1}{3}a^3 + a^3 - \tfrac{1}{3}a^3\right) \\
&= \tfrac{4}{3}\pi a b^2.
\end{aligned}
$$

Figure 11.8 shows an ellipsoid.
Sweeping the ellipse about the y-axis creates another ellipsoid, with a different
volume given by

$$
\begin{aligned}
V &= \pi \int_{-b}^{b} x^2 \, dy \\
&= \pi \frac{a^2}{b^2} \int_{-b}^{b} \left(b^2 - y^2\right) \, dy
\end{aligned}
$$

Fig. 11.8 An ellipsoid

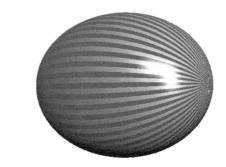

Fig. 11.9 A parabola, which
when rotated about the
y-axis creates a paraboloid

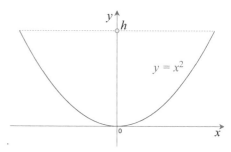

$$= \pi \frac{a^2}{b^2} \left[b^2 y - \tfrac{1}{3} y^3 \right]_{-b}^{b}$$

$$= \pi \frac{a^2}{b^2} \left(b^3 - \tfrac{1}{3} b^3 + b^3 - \tfrac{1}{3} b^3 \right)$$

$$= \tfrac{4}{3} \pi a^2 b.$$

Observe that in both cases when $a = b = r$, the object is a sphere with a volume of
$\tfrac{4}{3} \pi r^3$.

11.2.6 Volume of a Paraboloid

Figure 11.9 shows a parabola, which when rotated about the y-axis forms a 3D
paraboloid To rotate about the y-axis the equation of the parabola is

$$x = \sqrt{y}$$

where $y \in [0, h]$. The volume of the paraboloid is

$$V = \pi \int_0^h x^2 \, dy$$

Fig. 11.10 A paraboloid

Fig. 11.11 A series of concentric shells

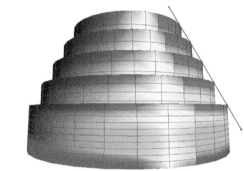

$$= \pi \int_0^h y \, dy$$

$$= \pi \cdot \frac{1}{2} y^2 \Big|_0^h$$

$$= \frac{1}{2} \pi h^2.$$

If $x \in [0, 1]$, then $h = 1$, and the volume is $\pi/2$. Figure 11.10 shows a paraboloid.

11.3 Solid of Revolution: Shells

A solid of revolution can also be constructed from a collection of concentric cylindrical shells as shown in Fig. 11.11, where the object's shape is defined by the contour $y = f(x)$ which is rotated about the y-axis. Figure 11.13 shows one of the cylindrical shells with a radius of x_i, $f(x_i)$ high and Δx thick. As the shell is assumed to be infinitesimally thin, the volume of the shell is

$$V_i = 2\pi x_i f(x_i) \Delta x.$$

Fig. 11.12 Dimensions for
one concentric shell

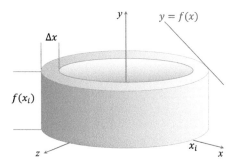

Dividing the solid into n such shells, and letting n tend towards infinity, the entire
volume is given by

$$V = \lim_{n \to \infty} \sum_{i=1}^{n} 2\pi x_i f(x_i) \, \Delta x$$

which in integral form is

$$V = 2\pi \int_a^b x \, f(x) \, dx. \tag{11.2}$$

Similarly, when the contour is rotated about the x-axis, the integral is

$$V = 2\pi \int_c^d y \, f(y) \, dy. \tag{11.3}$$

Let's test (11.2) and (11.3) with various contours.

11.3.1   Volume of a Cylinder

Figure 11.12 shows the geometry to create a cylinder with radius r, and height h to
be rotated about the y-axis. Using (11.2) the volume is

$$V = 2\pi \int_a^b x \, f(x) \, dx$$

$$= 2\pi \int_0^r xh \, dx$$

$$= 2\pi h \cdot \tfrac{1}{2} x^2 \Big|_0^r$$

$$= \pi r^2 h.$$

Fig. 11.13 The geometry used to create a cylinder

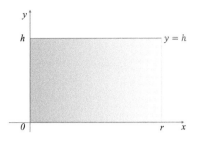

11.3.2 Volume of a Right Cone

Figure 11.14 shows a straight line represented by $y = h(1 - x/r)$, which when rotated about the y-axis sweeps out a right cone with radius r, and height h. Its volume is given by

$$V = 2\pi \int_0^r x\, f(x)\, dx$$

$$= 2\pi \int_0^r xh\left(1 - \frac{x}{r}\right) dx$$

$$= 2\pi h \int_0^r \left(x - \frac{x^2}{r}\right) dx$$

$$= 2\pi h \left[\tfrac{1}{2}x^2 - \tfrac{1}{3}\frac{x^3}{r}\right]_0^r$$

$$= 2\pi h \left(\tfrac{1}{2}r^2 - \tfrac{1}{3}r^2\right)$$

$$= \tfrac{1}{3}\pi r^2 h.$$

11.3.3 Volume of a Sphere

Figure 11.15 shows the geometry to create a hemisphere with radius r to be rotated about the y-axis. As we have seen before, it is convenient to use polar coordinates when dealing with circles and spheres, therefore, our equations are

$$x = r\cos\theta \quad \text{and} \quad y = r\sin\theta.$$

The original interval for x is $x \in [0, r]$, which for θ is $\theta \in [\pi/2, 0]$. Therefore,

$$\frac{dx}{d\theta} = -r\sin\theta \quad \text{or} \quad dx = -r\sin\theta\, d\theta.$$

Fig. 11.14 The geometry
used to create a right cone

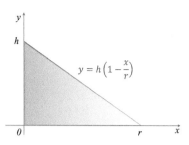

$$y = h\left(1 - \frac{x}{r}\right)$$

Fig. 11.15 The geometry
used to create a hemisphere

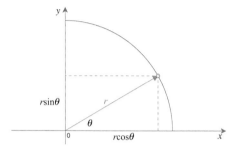

Using (11.2) the volume is

$$V = 2\pi \int_0^r x\, f(x)\, dx$$

$$= 2\pi \int_{\pi/2}^0 (r\cos\theta \cdot r\sin\theta\,(-r\sin\theta))\, d\theta$$

$$= -2\pi r^3 \int_{\pi/2}^0 \cos\theta \cdot \sin^2\theta\, d\theta$$

$$= -2\pi r^3 \int_{\pi/2}^0 \cos\theta\left(1 - \cos^2\theta\right) d\theta$$

$$= -2\pi r^3 \int_{\pi/2}^0 \cos\theta\, d\theta + 2\pi r^3 \int_{\pi/2}^0 \cos^3\theta\, d\theta$$

$$= -2\pi r^3 \cdot \sin\theta\,\Big|_{\pi/2}^0 + 2\pi r^3 \int_{\pi/2}^0 \cos^3\theta\, d\theta$$

$$= 2\pi r^3 + 2\pi r^3 \int_{\pi/2}^0 \cos^3\theta\, d\theta.$$

From Appendix B, we see that

$$\int \cos^3\theta\, d\theta = \tfrac{1}{3}\sin\theta \cdot \cos^2\theta + \tfrac{2}{3}\sin\theta + C.$$

Therefore,

$$V = 2\pi r^3 + 2\pi r^3 \left[\tfrac{1}{3} \sin \theta \cdot \cos^2 \theta + \tfrac{2}{3} \sin \theta \right]_{\pi/2}^{0}$$
$$= 2\pi r^3 - 2\pi r^3 \tfrac{2}{3}$$
$$= \tfrac{2}{3} \pi r^3$$

which makes a sphere's volume $\tfrac{4}{3}\pi r^3$.

11.3.4 Volume of a Paraboloid

We have already seen that the volume of a paraboloid using $y = x^2$ is $\tfrac{1}{2}\pi h^2$, where h is the height. The following shell method computes the volume surrounding the paraboloid, which using (11.2) gives

$$V = 2\pi \int_0^r x \, f(x) \, dx$$
$$= 2\pi \int_0^r x \cdot x^2 \, dx$$
$$= 2\pi \int_0^r x^3 \, dx$$
$$= 2\pi \cdot \tfrac{1}{4} x^4 \Big|_0^r$$
$$= \tfrac{1}{2} \pi r^4$$

and if $x \in [0, 1]$, then $h = r^2$, and $V = \tfrac{1}{2}\pi h^2$. Which shows that the volume of inner paraboloid equals the enclosing volume. In order to compute the volume of a paraboloid using the shell technique, the parabola has to be inverted, as shown in Fig. 11.16.

Fig. 11.16 The geometry used to create a paraboloid

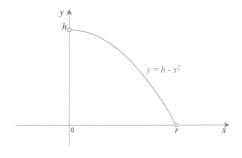

Fig. 11.17 A surface
created by $z = f(x, y)$

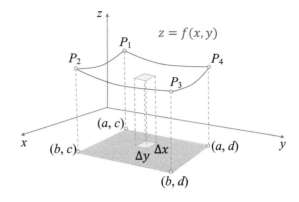

$$V = 2\pi \int_0^r x\, f(x)\, dx$$

$$= 2\pi \int_0^r x\, \left(h - x^2\right)\, dx$$

$$= 2\pi \int_0^r xh - x^3\, dx$$

$$= 2\pi \left[\tfrac{1}{2}x^2 h - \tfrac{1}{4}x^4\right]_0^r$$

$$= 2\pi \left(\tfrac{1}{2}r^2 h - \tfrac{1}{4}r^4\right).$$

But in our equation, $h = r^2$, therefore,

$$V = 2\pi \left(\tfrac{1}{2}h^2 - \tfrac{1}{4}h^2\right)$$

$$= \tfrac{1}{2}\pi h^2.$$

11.4   Volumes with Double Integrals

Figure 11.17 illustrates a 3D function where $z = f(x, y)$ over a region R defined
by the limits $a \leq x \leq b$ and $c \leq y \leq d$, whose area is projected onto the xy-plane.
If we consider a small rectangular tile on the xy-plane with dimensions Δx and Δy,
the volume of this column is approximately

$$\Delta V \approx f(x_i, y_j) \Delta x . \Delta y$$

where i and j identify a specific tile. Therefore, the total volume is

$$V \approx \sum_{i,j} f(x_i, y_j) \Delta x . \Delta y.$$

In the limit

$$V = \lim_{\Delta x, \Delta y \to 0} \sum_{i,j} f(x_i, y_j) \Delta x . \Delta y$$

or in integral form:

$$V = \int_a^b \int_c^d f(x, y) \, dx \, dy$$

where the inner integral is evaluated first, followed by the outer integral. The integral can be written in two ways:

$$V = \int_a^b \int_c^d f(x, y) \, dx \, dy = \int_c^d \int_a^b f(x, y) \, dy \, dx. \qquad (11.4)$$

Let's apply (11.4) in various scenarios.

11.4.1 Objects with a Rectangular Base

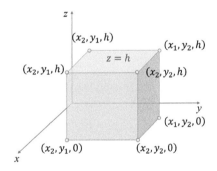

Fig. 11.18 A rectangular box

11.4.2 Rectangular Box

Figure 11.18 shows a rectangular box whose top surface is defined by $z = h$, with base dimensions $(x_2 - x_1)$ and $(y_2 - y_1)$, where the enclosed volume is

$$V = h(x_2 - x_1)(y_2 - y_1).$$

This is confirmed by (11.4) as follows:

$$V = \int_{y_1}^{y_2} \int_{x_1}^{x_2} f(x, y) \, dx \, dy$$

$$= \int_{y_1}^{y_2} \int_{x_1}^{x_2} h \, dx \, dy$$

$$= h \int_{y_1}^{y_2} \int_{x_1}^{x_2} 1 \, dx \, dy$$

$$= h \int_{y_1}^{y_2} x \Big|_{x_1}^{x_2} \, dy$$

$$= h \int_{y_1}^{y_2} (x_2 - x_1) \, dy$$

$$= h(x_2 - x_1) \int_{y_1}^{y_2} 1 \, dy$$

$$= h(x_2 - x_1) \cdot y \Big|_{y_1}^{y_2}$$

$$= h(x_2 - x_1)(y_2 - y_1).$$

11.4.3 Rectangular Prism

Figure 11.19 shows a rectangular prism whose top sloping surface is defined by $z = h(1 - x/a)$, with base dimensions a and b, where the enclosed volume is

$$V = \tfrac{1}{2}hab.$$

Fig. 11.19 A prism

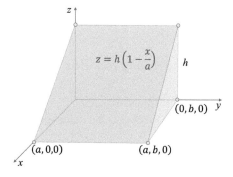

This is confirmed by (11.4) as follows:

$$
\begin{aligned}
V &= \int_{y_1}^{y_2} \int_{x_1}^{x_2} f(x, y) \, dx \, dy \\
&= \int_0^b \int_0^a h \left(1 - \frac{x}{a} \right) dx \, dy \\
&= h \int_0^b \int_0^a \left(1 - \frac{x}{a} \right) dx \, dy \\
&= h \int_0^b \left[x - \frac{1}{2} \frac{x^2}{a} \right]_0^a dy \\
&= h \int_0^b \left(a - \tfrac{1}{2}a \right) dy \\
&= \tfrac{1}{2}ha \int_0^b 1 \, dy \\
&= \tfrac{1}{2}ha \cdot y \Big|_0^b \\
&= \tfrac{1}{2}hab.
\end{aligned}
$$

11.4.4 Curved Top

Figure 11.20 shows an object with a square base and curved top defined by $z = x^2 + y$. Given that $\{x, y\} \in [0, 1]$, then the enclosed volume is:

Fig. 11.20 An object with a curved top

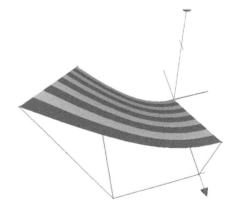

$$V = \int_{y_1}^{y_2} \int_{x_1}^{x_2} f(x, y) \, dx \, dy$$

$$= \int_0^1 \int_0^1 \left(x^2 + y\right) \, dx \, dy$$

$$= \int_0^1 \left[\tfrac{1}{3}x^3 + xy\right]_0^1 \, dy$$

$$= \int_0^1 \left(y + \tfrac{1}{3}\right) \, dy$$

$$= \left[\tfrac{1}{2}y^2 + \tfrac{1}{3}y\right]_0^1$$

$$= \tfrac{1}{2} + \tfrac{1}{3}$$

$$= \tfrac{5}{6}.$$

11.4.5 Objects with a Circular Base

The same double integral works with polar coordinates, which enables us to compute the volume of objects with a circular base. We have already seen that when moving from Cartesian coordinates to polar coordinates, the appropriate Jacobian must be included. In this case, the following substitutions are:

$$x = \rho \cos \theta$$
$$y = \rho \sin \theta$$
$$dx \, dy = \rho \, d\rho \, d\theta$$

which transforms (11.4) into

$$V = \int_a^b \int_c^d f(x, y) \, dx \, dy = \int_0^{2\pi} \int_0^r f(\rho \cos \theta, \rho \sin \theta) \rho \, d\rho \, d\theta. \qquad (11.5)$$

Let's test (11.5) using various objects.

11.4.6 Cylinder

The volume of a cylinder with radius r and $f(\rho \cos \theta, \rho \sin \theta) = h$ is $\pi r^2 h$, which is confirmed as follows:

Fig. 11.21 Cross section of
a cylinder and intersecting
plane

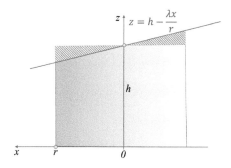

$$V = \int_0^{2\pi} \int_0^r f(\rho \cos \theta, \rho \sin \theta) \rho \, d\rho \, d\theta$$

$$= \int_0^{2\pi} \int_0^r h\rho \, d\rho \, d\theta$$

$$= h \int_0^{2\pi} \tfrac{1}{2}\rho^2 \Big|_0^r \, d\theta$$

$$= \tfrac{1}{2}r^2 h \int_0^{2\pi} 1 \, d\theta$$

$$= \tfrac{1}{2}r^2 h \cdot \theta \Big|_0^{2\pi}$$

$$= \pi r^2 h.$$

11.4.7 Truncated Cylinder

The volume of a truncated cylinder is calculated by forming the intersection of a cylinder and an oblique plane. The following proof confirms that the volume equals $\pi r^2 h$, because the cylinder's height, h, is the z-axis. To illustrate this, Fig. 11.21 shows a side projection of a cylinder intersecting the plane: $z = h - \lambda x/r$, where λ controls the slope of the plane. It is clear that the two cross-hatched triangles are equal, which is why the volume is unchanged:

$$V = \int_0^{2\pi} \int_0^r f(\rho \cos \theta, \rho \sin \theta) \rho \, d\rho \, d\theta$$

$$= \int_0^{2\pi} \int_0^r \left(h - \frac{\lambda \rho \cos \theta}{r} \right) \rho \, d\rho \, d\theta$$

$$= \int_0^{2\pi} \int_0^a \left(\rho h - \frac{\lambda \rho^2 \cos \theta}{r} \right) d\rho \, d\theta$$

Fig. 11.22 A cross-section of parabola intersecting a cylinder

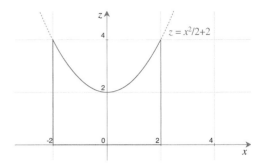

$$= \int_0^{2\pi} \left[\tfrac{1}{2}\rho^2 h - \frac{\lambda\rho^3 \cos\theta}{3r} \right]_0^r d\theta$$

$$= \int_0^{2\pi} \left(\tfrac{1}{2}r^2 h - \tfrac{1}{3}\lambda r^2 \cos\theta \right) d\theta$$

$$= \tfrac{1}{6}r^2 \int_0^{2\pi} (3h - 2\lambda \cos\theta) \, d\theta$$

$$= \tfrac{1}{6}r^2 \left[3h\theta - 2\lambda \sin\theta \right]_0^{2\pi}$$

$$= \tfrac{1}{6}r^2 6\pi h$$

$$= \pi r^2 h.$$

If the radius is 2, and the height 4, then the volume is 16π. Observe that the result is independent of λ. Taking this cylinder and intersecting it with the parabola, $z = 2 + \tfrac{1}{2}x^2$ as shown in Fig. 11.22, the volume reduces to 10π:

$$V = \int_0^{2\pi} \int_0^2 \left(2 + \tfrac{1}{2}x^2 \right) \rho \, d\rho \, d\theta$$

$$= \int_0^{2\pi} \int_0^2 \left(2 + \tfrac{1}{2}\rho^2 \cos^2\theta \right) \rho \, d\rho \, d\theta$$

$$= \int_0^{2\pi} \int_0^2 \left(2\rho + \tfrac{1}{2}\rho^3 \cos^2\theta \right) d\rho \, d\theta$$

$$= \int_0^{2\pi} \left[\rho^2 + \tfrac{1}{8}\rho^4 \cos^2\theta \right]_0^2 d\theta$$

$$= \int_0^{2\pi} 4 + 2\cos^2\theta \, d\theta$$

$$= \int_0^{2\pi} 5 + \cos 2\theta \, d\theta$$

$$= \left[5\theta + \tfrac{1}{2}\sin 2\theta \right]_0^{2\pi}$$

$$= 10\pi.$$

11.5   Volumes with Triple Integrals

The double integral for calculating area is

$$\iint_R f(x, y) \, dx \, dy \quad \text{or} \quad \iint_R f(x, y) \, dA$$

where the region R is divided into a matrix of small areas represented by $dx \, dy$ or dA. The Riemann sum notation is

$$\iint_R f(x, y) \, dA = \lim_{n \to \infty} \sum_{i=1}^{n} f(x_i, y_i) \, \Delta A_i.$$

This notation can be generalised into a triple integral for calculating volume:

$$\iiint_R f(x, y, z) \, dx \, dy \, dz \quad \text{or} \quad \iiint_R f(x, y, z) \, dV$$

where the region R is divided into a matrix of small volumes represented by $dx \, dy \, dz$ or dV. The Riemann sum notation is

$$\iint_R f(x, y, z) \, dV = \lim_{n \to \infty} \sum_{i=1}^{n} f(x_i, y_i, z_i) \, \Delta V_i.$$

Let's apply (11.6), where each integral identifies its interval of integration, to various 3D objects and calculate their volume.

$$V = \int_a^b \int_c^d \int_e^f f(x, y, z) \, dx \, dy \, dz. \tag{11.6}$$

11.5.1   Rectangular Box

Figure 11.23 shows the Cartesian coordinates for a rectangular box, with x-, y- and z-lengths are $(x_2 - x_1)$, $(y_2 - y_1)$ and $(z_2 - z_1)$ respectively, and whose volume is calculated using (11.6) as follows.

$$\begin{aligned}
V &= \int_a^b \int_c^d \int_e^f f(x, y, z) \, dx \, dy \, dz \\
&= \int_{z_1}^{z_2} \int_{y_1}^{y_2} \int_{x_1}^{x_2} 1 \, dx \, dy \, dz.
\end{aligned}$$

Fig. 11.23 Cartesian
coordinates for a rectangular
box

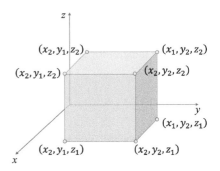

Together, the three integrals create the product of three lengths:

$$x_2 - x_1, \quad y_2 - y_1, \quad z_2 - z_1$$

which form the volume of the box:

$$
\begin{aligned}
V &= \int_{z_1}^{z_2} \int_{y_1}^{y_2} x \Big|_{x_1}^{x_2} \, dy \, dz \\
&= (x_2 - x_1) \int_{z_1}^{z_2} \int_{y_1}^{y_2} 1 \, dy \, dz \\
&= (x_2 - x_1) \int_{z_1}^{z_2} y \Big|_{y_1}^{y_2} \, dz \\
&= (x_2 - x_1)(y_2 - y_1) \int_{z_1}^{z_2} 1 \, dz \\
&= (x_2 - x_1)(y_2 - y_1) \cdot z \Big|_{z_1}^{z_2} \\
&= (x_2 - x_1)(y_2 - y_1)(z_2 - z_1)
\end{aligned}
$$

which confirms that the volume is the product of the box's linear measurements.

11.5.2   Volume of a Cylinder

Figure 11.24 shows a quadrant of a cylinder with radius r, and height h. Its volume
is computed by dividing the enclosed space into cuboids with a volume $\Delta V_i = \delta x \cdot \delta y \cdot \delta z$. In the limit, as δx, δy and δz tend towards zero, the entire volume is a
Riemann sum, and a triple integral:

$$V = \int_0^h \int_0^r \int_0^{\sqrt{r^2 - y^2}} 1 \, dx \, dy \, dz. \tag{11.7}$$

Fig. 11.24 The first
quadrant of a circular arc

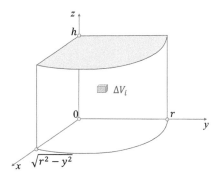

The solution looks neater if the integrals are evaluated as follows

$$V = \int_0^r \int_0^{\sqrt{r^2-y^2}} \int_0^h 1 \, dz \, dx \, dy$$

$$= \int_0^r \int_0^{\sqrt{r^2-y^2}} z \Big|_0^h \, dx \, dy$$

$$= h \int_0^r \int_0^{\sqrt{r^2-y^2}} 1 \, dx \, dy$$

$$= h \int_0^r x \Big|_0^{\sqrt{r^2-y^2}} \, dy$$

$$= h \int_0^r \sqrt{r^2 - y^2} \, dy.$$

Let $y = r \sin \theta$, then

$$\frac{dy}{d\theta} = r \cos \theta \quad \text{or} \quad dy = r \cos \theta \, d\theta$$

and the interval for θ is $\theta \in [0, \pi/2]$, therefore,

$$V = h \int_0^{\pi/2} \sqrt{r^2 - r^2 \sin^2 \theta} \cdot r \cos \theta \, d\theta$$

$$= r^2 h \int_0^{\pi/2} \cos^2 \theta \, d\theta$$

$$= \tfrac{1}{2} r^2 h \int_0^{\pi/2} (1 + \cos 2\theta) \, d\theta$$

$$= \tfrac{1}{2} r^2 h \Big[\theta + \tfrac{1}{2} \sin 2\theta \Big]_0^{\pi/2}$$

$$= \tfrac{1}{4} \pi r^2 h.$$

As there are four such quadrants, the cylinder's volume is $\pi r^2 h$.

Cartesian coordinates are not best suited for this work—it is much more convenient to employ cylindrical polar coordinates, where

$$x = \rho \cos \phi, \quad y = \rho \sin \phi, \quad z = z$$

and the Jacobian is ρ. Therefore, (11.7) is written to represent the entire volume as

$$V = \int_0^h \int_0^{2\pi} \int_0^r \rho \, d\rho \, d\phi \, dz$$

which is integrated as follows:

$$V = \int_0^h \int_0^{2\pi} \int_0^r \rho \, d\rho \, d\phi \, dz$$
$$= \int_0^h \int_0^{2\pi} \tfrac{1}{2}\rho^2 \Big|_0^r \, d\phi \, dz$$
$$= \tfrac{1}{2}r^2 \int_0^h \int_0^{2\pi} 1 \, d\phi \, dz$$
$$= \tfrac{1}{2}r^2 \int_0^h \phi \Big|_0^{2\pi} \, dz$$
$$= \pi r^2 \int_0^h 1 \, dz$$
$$= \pi r^2 \cdot z \Big|_0^h$$
$$= \pi r^2 h.$$

11.5.3   Volume of a Sphere

Figure 11.25 shows how a sphere is defined using spherical polar coordinates, where any point has the coordinates (ρ, ϕ, θ). In order to compute its volume, the following intervals apply: $\rho \in [0, r]$, $\phi \in [0, \pi]$, and $\theta \in [0, 2\pi]$. Using the Jacobian $\rho^2 \sin \phi$, the volume is

$$V = \int_0^{2\pi} \int_0^{\pi} \int_0^r \rho^2 \sin \phi \, d\rho \, d\phi \, d\theta$$
$$= \int_0^{2\pi} \int_0^{\pi} \tfrac{1}{3}\rho^3 \Big|_0^r \sin \phi \, d\phi \, d\theta$$
$$= \tfrac{1}{3}r^3 \int_0^{2\pi} \int_0^{\pi} \sin \phi \, d\phi \, d\theta$$
$$= \tfrac{1}{3}r^3 \int_0^{2\pi} -\cos \phi \Big|_0^{\pi} \, d\theta$$

Fig. 11.25 Spherical polar coordinates

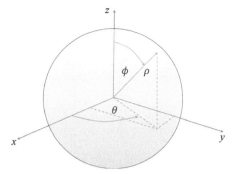

Fig. 11.26 A cone with cylindrical coordinates

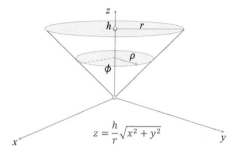

$$= \tfrac{2}{3}r^3 \int_0^{2\pi} 1 \, d\theta$$

$$= \tfrac{2}{3}r^3 \cdot \theta \Big|_0^{2\pi}$$

$$= \tfrac{4}{3}\pi r^3.$$

11.5.4 *Volume of a Cone*

The triple integral provides another way to compute the volume of a cone, and is best evaluated using cylindrical polar coordinates, rather than Cartesian coordinates. Figure 11.26 shows an inverted cone with height h and radius r. The equation for the cone is given by

$$z = \frac{h}{r}\sqrt{x^2 + y^2}$$

where any point in the cone has a distance $\rho = \sqrt{x^2 + y^2}$ from the z-axis. Thus when $\rho = r, z = h$, and when $\rho = 0, z = 0$, which provides the cone's shape. We are only interested in the volume between $z = 0$ and $z = h$.

Thus the intervals for the three cylindrical coordinates are: $\phi \in [0, 2\pi]$, $\rho \in [0, r]$ and $z \in \left[\frac{h}{r}\rho, h\right]$, and using the Jacobian ρ, the triple integral is

$$V = \int_0^r \int_0^{2\pi} \int_{h\rho/r}^h \, d\phi \, dz \, \rho \, d\rho.$$

Integrating from the inside outwards, we have

$$V = \int_0^r \int_{h\rho/r}^h \int_0^{2\pi} \, d\phi \, dz \, \rho \, d\rho$$

$$= \int_0^r \int_{h\rho/r}^h \int_0^{2\pi} \, d\phi \, dz \, \rho \, d\rho$$

$$= \int_0^r \int_{h\rho/r}^h \phi \Big|_0^{2\pi} \, dz \, \rho \, d\rho$$

$$= 2\pi \int_0^r \int_{h\rho/r}^h 1 \, dz \, \rho \, d\rho$$

$$= 2\pi \int_0^r z \Big|_{h\rho/r}^h \, \rho \, d\rho$$

$$= 2\pi \int_0^r \left(h - \frac{h\rho}{r}\right) \rho \, d\rho$$

$$= 2\pi \int_0^r \left(h\rho - \frac{h\rho^2}{r}\right) d\rho$$

$$= 2\pi \left[\tfrac{1}{2}h\rho^2 - \frac{h\rho^3}{3r}\right]_0^r$$

$$= 2\pi \left(\tfrac{1}{2}hr^2 - \tfrac{1}{3}hr^2\right)$$

$$= \frac{2\pi}{6} \left(3hr^2 - 2hr^2\right)$$

$$= \tfrac{1}{3}\pi hr^2.$$

11.6   Summary

Integral Calculus is a powerful tool for computing volume, whether it be using single, double or triple integrals, and this chapter has covered four techniques using the following formulae.

11.6.1   Summary of Formulae

Slicing: Rotating $f(x)$ about the x-axis

$$V = \pi \int_a^b (f(x))^2 \ dx.$$

Slicing: Rotating $f(y)$ about the y-axis

$$V = \pi \int_a^b (f(y))^2 \ dy.$$

Shells: Rotating $f(x)$ about the x-axis

$$V = 2\pi \int_a^b x \ f(x) \ dx.$$

Shells: Rotating $f(x)$ about the y-axis

$$V = 2\pi \int_a^b y \ f(y) \ dy.$$

Surface function $f(x, y)$ using rectangular coordinates

$$V = \int_a^b \int_c^d f(x, y) \ dx \ dy = \int_c^d \int_a^b f(x, y) \ dy \ dx.$$

Surface function $f(x, y)$ using polar coordinates

$$V = \int_a^b \int_c^d f(x, y) \ dx \ dy = \int_{\rho_{min}}^{\rho_{max}} \int_{\theta_{min}}^{\theta_{max}} f(\rho \cos\theta, \rho \sin\theta) \ \rho \ d\theta \ d\rho.$$

Triple integral using rectangular coordinates

$$V = \int_a^b \int_c^d \int_e^f f(x, y, z) \ dx \ dy \ dz.$$

Triple integral using cylindrical polar coordinates

$$V = \int_{z_{min}}^{z_{max}} \int_{\phi_{min}}^{\phi_{max}} \int_{\rho_{min}}^{\rho_{max}} f(\rho, \phi, z) \ \rho \ d\rho \ d\phi \ dz.$$

Chapter 12
Vector-Valued Functions

12.1 Introduction

So far, all the functions we have differentiated or integrated have been real-valued functions, such as

$$f(x) = x + \sin x$$

where x is a real value. However, as vectors play such an important role in physics, mechanics, motion, etc., it is essential that we understand how to differentiate and integrate vector-valued functions such as

$$\mathbf{p}(t) = x(t)\mathbf{i} + y(t)\mathbf{j} + z(t)\mathbf{k}$$

where $\mathbf{i}$, $\mathbf{j}$ and $\mathbf{k}$ are unit basis vectors. This chapter introduces how such functions are differentiated and integrated.

12.2 Differentiating Vector Functions

The position of a point $P(x, y)$ on the plane is located using a vector:

$$\mathbf{p} = x\mathbf{i} + y\mathbf{j}$$

or a point $P(x, y, z)$ in 3D space as

$$\mathbf{p} = x\mathbf{i} + y\mathbf{j} + z\mathbf{k}.$$

© Springer Nature Switzerland AG 2019
J. Vince, *Calculus for Computer Graphics*,
https://doi.org/10.1007/978-3-030-11376-6_12

If the point is moving and controlled by a time-based function with parameter t, then the position vector has the form:

$$\mathbf{p}(t) = x(t)\mathbf{i} + y(t)\mathbf{j}$$

or in 3D space

$$\mathbf{p}(t) = x(t)\mathbf{i} + y(t)\mathbf{j} + z(t)\mathbf{k}.$$

The derivative of $\mathbf{p}(t)$ is another vector formed from the derivatives of $x(t)$, $y(t)$ and $z(t)$:

$$\frac{d}{dt}\mathbf{p}(t) = \mathbf{p}'(t) = \frac{dx}{dt}\mathbf{i} + \frac{dy}{dt}\mathbf{j}$$

or in 3D:

$$\frac{d}{dt}\mathbf{p}(t) = \mathbf{p}'(t) = \frac{dx}{dt}\mathbf{i} + \frac{dy}{dt}\mathbf{j} + \frac{dz}{dt}\mathbf{k}.$$

For example, given

$$\mathbf{p}(t) = 10\sin t\mathbf{i} + 5t^2\mathbf{j} + 20\cos t\mathbf{k}$$

then

$$\frac{d}{dt}\mathbf{p}(t) = 10\cos t\mathbf{i} + 10t\mathbf{j} - 20\sin t\mathbf{k}.$$

12.2.1   Velocity and Speed

As $\mathbf{p}(t)$ gives the position of a point at time t, its derivative gives the rate of change of the position with respect to time, i.e. its velocity. For example, if $\mathbf{p}(t)$ is the position of a point P at time t, P's change in position from t to $t + \Delta t$ is

$$\Delta\mathbf{p} = \mathbf{p}(t + \Delta t) - \mathbf{p}(t).$$

Dividing throughout by Δt:

$$\frac{\Delta\mathbf{p}}{\Delta t} = \frac{\mathbf{p}(t + \Delta t) - \mathbf{p}(t)}{\Delta t}.$$

In the limit as $\Delta t \to 0$ we have

$$\frac{d}{dt}\mathbf{p}(t) = \mathbf{v}(t) = \lim_{\Delta t \to 0}\frac{\mathbf{p}(t + \Delta t) - \mathbf{p}(t)}{\Delta t}$$

which is the velocity of P at time t. Figure 12.2 shows this diagrammatically.

Fig. 12.1 Velocity of P at time t

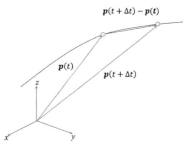

Fig. 12.2 Position and velocity vectors for P

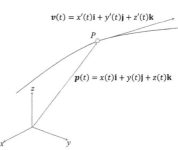

For example, if the functions controlling a particle are $x(t) = 3\cos t$, $y(t) = 4\sin t$ and $z(t) = 5t$, then

$$\mathbf{p}(t) = 3\cos t\,\mathbf{i} + 4\sin t\,\mathbf{j} + 5t\,\mathbf{k}$$

and differentiating $\mathbf{p}(t)$ gives the velocity vector:

$$\mathbf{v}(t) = -3\sin t\,\mathbf{i} + 4\cos t\,\mathbf{j} + 5\mathbf{k}.$$

Figure 12.2 shows a point P moving along a trajectory defined by its position vector $\mathbf{p}(t)$. P's velocity is represented by $\mathbf{v}(t)$ which is tangential to the trajectory at P (Fig. 12.2).

Given the position vector for a particle P,

$$\mathbf{p}(t) = x(t)\mathbf{i} + y(t)\mathbf{j} + z(t)\mathbf{k}$$

the speed of P is given by

$$|\mathbf{v}(t)| = \sqrt{\left(\frac{dx}{dt}\right)^2 + \left(\frac{dy}{dt}\right)^2 + \left(\frac{dz}{dt}\right)^2}.$$

In the case of

$$\mathbf{v}(t) = -3\sin t\,\mathbf{i} + 4\cos t\,\mathbf{j} + 5\mathbf{k}.$$

the speed is

$$|\mathbf{v}(t)| = \sqrt{(-3\sin t)^2 + (4\cos t)^2 + 5^2}$$
$$= \sqrt{9\sin^2 t + 16\cos^2 t + 25}$$

and at time $t = 0$

$$|\mathbf{v}(t)| = \sqrt{16 + 25} = \sqrt{41}$$

and at time $t = \pi/2$

$$|\mathbf{v}(t)| = \sqrt{9 + 25} = \sqrt{34}.$$

12.2.2   Acceleration

The acceleration of a particle with position vector $\mathbf{p}(t)$ is the second derivative of $\mathbf{p}(t)$, or the derivative of P's velocity vector:

$$\mathbf{a}(t) = \mathbf{p}''(t) = \mathbf{v}'(t) = \frac{d^2x}{dt^2}\mathbf{i} + \frac{d^2y}{dt^2}\mathbf{j} + \frac{d^2z}{dt^2}\mathbf{k}.$$

In the case of

$$\mathbf{p}(t) = 3\cos t\mathbf{i} + 4\sin t\mathbf{j} + 5t\mathbf{k}$$
$$\mathbf{v}(t) = -3\sin t\mathbf{i} + 4\cos t\mathbf{j} + 5\mathbf{k}$$
$$\mathbf{a}(t) = -3\cos t\mathbf{i} - 4\sin t\mathbf{j}.$$

12.2.3   Rules for Differentiating Vector-Valued Functions

Vector-valued functions are treated just like vectors, in that they can be added, subtracted, scaled and multiplied, which leads to the following rules for their differentiation:

$$\frac{d}{dt}\left(\mathbf{p}(t) \pm \mathbf{q}(t)\right) = \frac{d}{dt}\mathbf{p}(t) \pm \frac{d}{dt}\mathbf{q}(t) \quad \text{addition and subtraction}$$

$$\frac{d}{dt}\left(\lambda\mathbf{p}(t)\right) = \lambda\frac{d}{dt}\mathbf{p}(t) \quad \text{where } \lambda \in R, \quad \text{scalar multiplier}$$

$$\frac{d}{dt}\left(f(t)\mathbf{p}(t)\right) = f(t)\mathbf{p}'(t) + f'(t)\mathbf{p}(t) \quad \text{function multiplier}$$

$$\frac{d}{dt}\big(\mathbf{p}(t) \cdot \mathbf{q}(t)\big) = \mathbf{p}(t) \cdot \mathbf{q}'(t) + \mathbf{p}'(t) \cdot \mathbf{q}(t) \quad \text{dot product}$$

$$\frac{d}{dt}\big(\mathbf{p}(t) \times \mathbf{q}(t)\big) = \mathbf{p}(t) \times \mathbf{q}'(t) + \mathbf{p}'(t) \times \mathbf{q}(t) \quad \text{cross product}$$

$$\frac{d}{dt}\big(\mathbf{p}(f(t))\big) = \mathbf{p}'\big(f(t)\big) f'(t) \quad \text{function of a function.}$$

12.3 Integrating Vector-Valued Functions

The integral of a vector-valued function is just its antiderivative, where each term is integrated individually. For example, given

$$\mathbf{p}(t) = x(t)\mathbf{i} + y(t)\mathbf{i} + z(t)\mathbf{k}$$

then

$$\int_a^b \mathbf{p}(t)\, dt = \int_a^b x(t)\mathbf{i}\, dt + \int_a^b y(t)\mathbf{i}\, dt + \int_a^b z(t)\mathbf{k}\, dt.$$

Similarly,

$$\int \mathbf{p}(t)\, dt = \int x(t)\mathbf{i}\, dt + \int y(t)\mathbf{i}\, dt + \int z(t)\mathbf{k}\, dt + \mathbf{C}.$$

Integrating the velocity vector used before:

$$\mathbf{v}(t) = -3 \sin t\mathbf{i} + 4 \cos t\mathbf{j} + 5\mathbf{k}$$

then

$$\int \mathbf{v}(t)\, dt = \int -3 \sin t\mathbf{i}\, dt + \int 4 \cos t\mathbf{j}\, dt + \int 5\mathbf{k}\, dt + \mathbf{C}$$

$$= -3 \int \sin t\mathbf{i}\, dt + 4 \int \cos t\mathbf{j}\, dt + 5 \int 1\mathbf{k}\, dt + \mathbf{C}$$

$$= 3 \cos t\mathbf{i} + 4 \sin t\mathbf{j} + 5t\mathbf{k} + \mathbf{C}.$$

We have already seen that

$$\mathbf{v}(t) = \frac{d}{dt}\mathbf{p}(t)$$

$$\mathbf{a}(t) = \frac{d}{dt}\mathbf{v}(t)$$

therefore,

$$\mathbf{p}(t) = \int \mathbf{v}(t) \, dt$$

$$\mathbf{v}(t) = \int \mathbf{a}(t) \, dt.$$

12.3.1   Velocity of a Falling Object

If an object falls under the influence of gravity ($9.8 \, \text{m/s}^2$) for $3 \, \text{s}$, its velocity at any time is given by

$$\mathbf{v}(t) = \int 9.8 \, dt = 9.8t + C_1.$$

Assuming that its initial velocity is zero, then $\mathbf{v}(0) = 0$, and $C_1 = 0$. Therefore,

$$\mathbf{p}(t) = \int 9.8t \, dt = \tfrac{9.8}{2}t^2 + C_2 = 4.9t^2 + C_2.$$

But $\mathbf{p}(0) = 0$, and $C_2 = 0$, therefore,

$$\mathbf{p}(t) = 4.9t^2.$$

Consequently, after $3 \, \text{s}$, the object has fallen $4.9 \times 3^2 = 40.1$ m.

   If the object had been given an initial downward velocity of $1 \, \text{m/s}$, then $C_1 = 1$, which means that

$$\mathbf{p}(t) = \int 9.8t + 1 \, dt = \tfrac{9.8}{2}t^2 + t + C_2 = 4.9t^2 + t + C_2.$$

But $\mathbf{p}(0) = 0$, and $C_2 = 0$, therefore,

$$\mathbf{p}(t) = 4.9t^2 + t.$$

Consequently, after $3 \, \text{s}$, the object has fallen $4.9 \times 3^2 + 3 = 43.1$ m.

12.3.2   Position of a Moving Object

Let's compute an object's position after $2 \, \text{s}$ if it is following a parametric curve such that its velocity is

$$\mathbf{v}(t) = t^2 \mathbf{i} + t \mathbf{j} + t^3 \mathbf{k}$$

starting at the origin at time $t = 0$.

$$\mathbf{p}(t) = \int \mathbf{v}(t)\, dt + \mathbf{C}$$

$$= \int t^2\mathbf{i} + t\mathbf{j} + t^3\mathbf{k}\, dt + \mathbf{C}$$

$$= \int t^2\mathbf{i}\, dt + \int t\mathbf{j}\, dt + \int t^3\mathbf{k}\, dt + \mathbf{C}$$

$$= \tfrac{1}{3}t^3\mathbf{i} + \tfrac{1}{2}t^2\mathbf{j} + \tfrac{1}{4}t^4\mathbf{k} + \mathbf{C}.$$

But $\mathbf{p}(0) = 0\mathbf{i} + 0\mathbf{j} + 0\mathbf{k}$, therefore, the vector $\mathbf{C} = 0\mathbf{i} + 0\mathbf{j} + 0\mathbf{k}$, and

$$\mathbf{p}(t) = \tfrac{1}{3}t^3\mathbf{i} + \tfrac{1}{2}t^2\mathbf{j} + \tfrac{1}{4}t^4\mathbf{k}.$$

Consequently, after 2 s, the object is at

$$\mathbf{p}(2) = \tfrac{1}{3}2^3\mathbf{i} + \tfrac{1}{2}2^2\mathbf{j} + \tfrac{1}{4}2^4\mathbf{k}$$

$$= \tfrac{8}{3}\mathbf{i} + 2\mathbf{j} + 4\mathbf{k}$$

which is the point $(8/3, 2, 4)$.

12.4 Summary

The Calculus of vector-based functions is a large and complex subject, and in this short chapter we have only covered the basic principles for differentiating and integrating simple functions, which are summarised next.

12.4.1 Summary of Formulae

Given a function of the form

$$\mathbf{p}(t) = x(t)\mathbf{i} + y(t)\mathbf{j} + z(t)\mathbf{k}$$

its derivative is

$$\frac{d}{dt}\mathbf{p}(t) = \mathbf{p}'(t) = \frac{dx}{dt}\mathbf{i} + \frac{dy}{dt}\mathbf{j} + \frac{dz}{dt}\mathbf{k}$$

its integral is

$$\int \mathbf{p}(t)\, dt = \int x(t)\mathbf{i}\, dt + \int y(t)\mathbf{i}\, dt + \int z(t)\mathbf{k}\, dt + \mathbf{C}$$

and definite integral:

$$\int_a^b \mathbf{p}(t)\, dt = \int_a^b x(t)\mathbf{i}\, dt + \int_a^b y(t)\mathbf{i}\, dt + \int_a^b z(t)\mathbf{k}\, dt.$$

If $\mathbf{p}(t)$ is a time-based position vector, its derivative is a velocity vector, and its second derivative is an acceleration vector:

$$\mathbf{p}(t) = x(t)\mathbf{i} + y(t)\mathbf{j} + z(t)\mathbf{k}$$
$$\mathbf{v}(t) = \frac{dx}{dt}\mathbf{i} + \frac{dy}{dt}\mathbf{j} + \frac{dz}{dt}\mathbf{k}$$
$$\mathbf{a}(t) = \frac{d^2x}{dt^2}\mathbf{i} + \frac{d^2y}{dt^2}\mathbf{j} + \frac{d^2z}{dt^2}\mathbf{k}.$$

The magnitude of $\mathbf{v}(t)$ represents speed:

$$|\mathbf{v}(t)| = \sqrt{\left(\frac{dx}{dt}\right)^2 + \left(\frac{dy}{dt}\right)^2 + \left(\frac{dz}{dt}\right)^2}$$

and for acceleration:

$$|\mathbf{a}(t)| = \sqrt{\left(\frac{d^2x}{dt^2}\right)^2 + \left(\frac{d^2y}{dt^2}\right)^2 + \left(\frac{d^2z}{dt^2}\right)^2}.$$

Chapter 13
Tangent and Normal Vectors

13.1 Introduction

In this chapter I describe how to calculate tangent and normal vectors on various curves and surfaces. I begin with the notation used to describe vector-valued functions and definitions for a tangent and normal vector. This includes an introduction to the grad operator, and how it is used to compute the gradient of a scalar field. I then show how these vectors are computed for a line, parabola, circle, ellipse, sine curve, cosh curve, helix, Bézier curve, bilinear patch, quadratic Bézier patch, sphere and a torus.

13.2 Notation

The following chapters refer to many vector-valued parametric functions, for which there are three popular forms of notation. The first employs a row vector:

$$\mathbf{r}(t) = [x(t) \ \ y(t) \ \ z(t)],$$

the second, a column vector:

$$\mathbf{r}(t) = \begin{bmatrix} x(t) \\ y(t) \\ z(t) \end{bmatrix},$$

and the third, a Cartesian vector:

$$\mathbf{r}(t) = x(t)\mathbf{i} + y(t)\mathbf{j} + z(t)\mathbf{k}.$$

I will tend to use column vectors and Cartesian notation to describe vector-valued functions.

© Springer Nature Switzerland AG 2019
J. Vince, *Calculus for Computer Graphics*,
https://doi.org/10.1007/978-3-030-11376-6_13

Fig. 13.1 The graphs of
$y = x^3$, (blue) and $y' = 3x^2$,
(green) its derivative

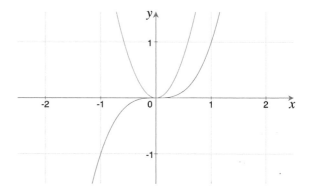

13.3   Tangent Vector to a Curve

We know that the derivative of a function measures the rate of change of the function
with respect to some parameter. In terms of the function's graph, the derivative is the
slope of the graph at a point. For instance, the function $y(x) = x^3$, the first derivative
is $y'(x) = 3x^2$, as shown in Fig. 13.1. The derivative is also the slope of the *tangent
vector*, whose magnitude and direction depend upon the form of parameterisation
used for the function. For example, defining a cubic as

$$\mathbf{r}(t) = t\mathbf{i} + t^3\mathbf{j}$$

the tangent vector is

$$\frac{d\mathbf{r}}{dt} = \mathbf{r}'(t) = \mathbf{i} + 3t^2\mathbf{j}$$

whose magnitude is

$$||\mathbf{r}'(t)|| = \sqrt{(1)^2 + (3t^2)^2} = \sqrt{1 + 9t^4}.$$

Figure 13.2 shows the cubic curve, with five tangent vectors for $t = -0.75, -0.5$,
$0.0, 0.5, 0.75$, which reflect the slope of the curve at the five points. However, in
definitions for curvature, a *unit tangent vector* is important, which requires dividing
the tangent vector by its magnitude:

$$\mathbf{T}(t) = \frac{\mathbf{r}'(t)}{||\mathbf{r}'(t)||}.$$

$\mathbf{T}(t)$ is defined, only if $\mathbf{r}'(t) \neq \mathbf{0}$.

The rate of change of the unit tangent vector gives the curvature $\kappa(t)$ at any point
along the curve length s:

Fig. 13.2 The graph of
$y = x^3$, and five tangent
vectors

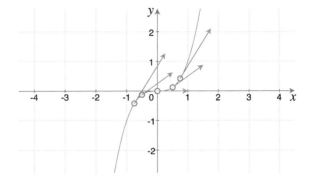

Fig. 13.3 The graph of
$y = x^3$, and five unit tangent
vectors

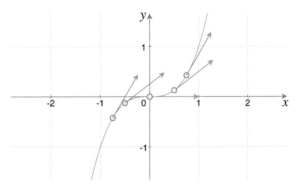

$$\kappa(t) = \frac{d\mathbf{T}}{ds}$$

which is covered in detail in Chap. 15. Figure 13.3 shows the cubic curve with five
unit tangent vectors.

Generally, for a vector-valued function $\mathbf{r}(t)$, that is continuously differentiable

$$\mathbf{r}(t) = \begin{bmatrix} x(t) \\ y(t) \end{bmatrix} \in \mathbb{R}^2, \qquad \mathbf{r}(t) = \begin{bmatrix} x(t) \\ y(t) \\ z(t) \end{bmatrix} \in \mathbb{R}^3$$

its tangent vector is

$$\frac{d\mathbf{r}}{dt} = \mathbf{r}'(t) = \begin{bmatrix} x'(t) \\ y'(t) \end{bmatrix} \neq \mathbf{0}, \qquad \frac{d\mathbf{r}}{dt} = \mathbf{r}'(t) = \begin{bmatrix} x'(t) \\ y'(t) \\ z'(t) \end{bmatrix} \neq \mathbf{0}.$$

For example, a constant pitch helix with radius ρ, is defined as

$$\mathbf{r}(t) = \begin{bmatrix} \rho \cos t \\ \rho \sin t \\ ct \end{bmatrix} = \rho \cos t\mathbf{i} + \rho \sin t\mathbf{j} + ct\mathbf{k}$$

therefore, its tangent vector is

$$\mathbf{r}'(t) = \begin{bmatrix} -\rho \sin t \\ \rho \cos t \\ c \end{bmatrix} = -\rho \sin t\mathbf{i} + \rho \cos t\mathbf{j} + c\mathbf{k}.$$

13.4   Normal Vector to a Curve

Ideally, a *normal vector* is orthogonal to a curve or surface, and orthogonal to its associated tangent vector. However, it would useful to confirm this mathematically. Once again, we are interested in the unit form, denoted by $\mathbf{N}(t)$.

By definition:

$$\|\mathbf{T}(t)\| = 1$$

therefore,

$$\|\mathbf{T}(t)\|^2 = 1$$

and as the dot product $\mathbf{T}(t) \cdot \mathbf{T}(t) = 1$

$$\|\mathbf{T}(t)\|^2 = \mathbf{T}(t) \cdot \mathbf{T}(t) = 1 \tag{13.1}$$

Differentiating (13.1), and bearing in mind that the dot product is commutative, we get

$$\frac{d}{dt}[\mathbf{T}(t) \cdot \mathbf{T}(t)] = \mathbf{T}'(t) \cdot \mathbf{T}(t) + \mathbf{T} \cdot \mathbf{T}'(t)$$

$$= 2\mathbf{T}'(t) \cdot \mathbf{T}(t) = 0.$$

For $\mathbf{T}'(t) \cdot \mathbf{T}(t) = 0$, $\mathbf{T}'(t)$ must be orthogonal to $\mathbf{T}(t)$, or $\mathbf{T}'(t) = 0$.

Thus we can define $\mathbf{N}(t)$ as

$$\mathbf{N}(t) = \frac{\mathbf{T}'(t)}{\|\mathbf{T}'(t)\|}.$$

Also, given a tangent vector $\mathbf{T}(t)$:

$$\mathbf{T}(t) = \begin{bmatrix} \lambda_1 \\ \lambda_2 \end{bmatrix} = \lambda_1\mathbf{i} + \lambda_2\mathbf{j}$$

Fig. 13.4 The graph of
$y = x^3$, with unit tangent
vectors (green), and unit
normal vectors (red)

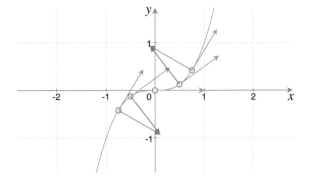

then two vectors exist, perpendicular to $\mathbf{T}(t)$:

$$\mathbf{N}_a = \begin{bmatrix} -\lambda_2 \\ \lambda_1 \end{bmatrix} = -\lambda_2\mathbf{i} + \lambda_1\mathbf{j} \quad \text{or} \quad \mathbf{N}_b = \begin{bmatrix} \lambda_2 \\ -\lambda_1 \end{bmatrix} = \lambda_2\mathbf{i} - \lambda_1\mathbf{j}$$

as the dot product $\mathbf{N}_a \cdot \mathbf{T}(t) = \mathbf{N}_b \cdot \mathbf{T}(t) = 0$, which means that $\mathbf{N}_a$ and $\mathbf{N}_b$ are normal vectors. Furthermore, if $\mathbf{T}(t)$ is a unit vector, so too, are $\mathbf{N}_a$ and $\mathbf{N}_b$.

But which one should we choose? Figure 13.4 shows a convention, where we see the unit normal vectors directed towards the zone containing the centre of curvature. This is called the *principal normal vector*. Another convention is to place the normal vector on one's right-hand side whilst traversing the curve.

You will notice from Fig. 13.4 that there is no normal vector when $t = 0$. Let's see why.

$$\mathbf{r}'(t) = \mathbf{i} + 3t^2\mathbf{j}$$
$$\mathbf{r}'(0) = \mathbf{i}$$
$$\|\mathbf{r}'(t)\| = \sqrt{1 + 9t^4}$$
$$\|\mathbf{r}'(0)\| = 1$$
$$\mathbf{T}(0) = \mathbf{i}$$
$$\mathbf{T}'(0) = 0.$$

So here is a case when $\mathbf{T}'(t) = 0$.

13.5   Gradient of a Scalar Field

The *grad* operator ∇, called 'del' or 'nabla' is a very useful idea for computing normal vectors within scalar fields. So first, let's define a scalar field.

Fig. 13.5 A scalar field for $f(x, y) = x^2 + y^2$

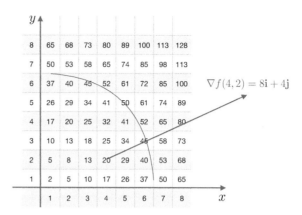

For our purpose, a scalar field is a 2D or 3D space where any point is represented by a scalar value, such as height, temperature or the value of a geometric function. On the other hand, any point in a vector field is represented by a vector, rather than a scalar. Figure 13.5 shows an array of scalars that obey the rule $f(x, y) = x^2 + y^2$. For example, the cell with (x, y) coordinates $(5, 5)$ contains 50, because $50 = 5^2 + 5^2$. In reality, a scalar field is continuous, rather than discrete, as shown in Fig. 13.5.

One can see from Fig. 13.5 that this scalar field comprises a family of concentric contours, one of which is sketched in the figure.

Taking the partial derivative of $f(x, y) = x^2 + y^2$ in the x-direction:

$$\frac{\partial f}{\partial x} = 2x$$

gives the instantaneous rate of change at any point (x, y) irrespective of the value of y. Similarly, taking the partial derivative of $f(x, y) = x^2 + y^2$ in the y-direction:

$$\frac{\partial f}{\partial y} = 2y$$

gives the instantaneous rate of change at any point (x, y) irrespective of the value of x.

The grad operator maps a scalar field to a vector field using these partial derivatives as follows:

$$\nabla f(x, y) = \begin{bmatrix} \frac{\partial f}{\partial x} \\ \frac{\partial f}{\partial y} \end{bmatrix} = \frac{\partial f}{\partial x}\mathbf{i} + \frac{\partial f}{\partial y}\mathbf{j}$$
$$= 2x\mathbf{i} + 2y\mathbf{j}.$$

For example, when $x = 4$ and $y = 2$:

$$\nabla f(4, 2) = 8\mathbf{i} + 4\mathbf{j}$$

which is sketched in Fig. 13.5. Observe that the vector is orthogonal to the contour. In fact, all vectors are orthogonal to all such contours. In other words, the vector is normal to any curve defined by the function $f(x, y) = x^2 + y^2$. For example, the equation of a circle is

$$x^2 + y^2 = r^2$$

where r is the radius. Therefore, we can create a function

$$f(x, y) = x^2 + y^2 - r^2 = 0.$$

Therefore,

$$\nabla f = 2x\mathbf{i} + 2y\mathbf{j}$$

which is the normal vector at (x, y), as shown in Fig. 13.5. To create a unit normal vector, we divide by the magnitude of the vector:

$$\mathbf{N} = \frac{\nabla f}{\|\nabla f\|}$$

For functions with three variables, the grad operator creates a 3D vector. For example:

$$f(x, y, z) = 2xy + 3z$$
$$\frac{\partial f}{\partial x} = 2y$$
$$\frac{\partial f}{\partial y} = 2x$$
$$\frac{\partial f}{\partial z} = 3$$
$$\nabla f = 2y\mathbf{i} + 2x\mathbf{j} + 3\mathbf{k}$$

which is normal to a surface defined by $f(x, y, z) = 2xy + 3z$.

I will employ the grad operator to create a normal vector to some of the following curves and surfaces.

Fig. 13.6 Geometry for a
parametric line

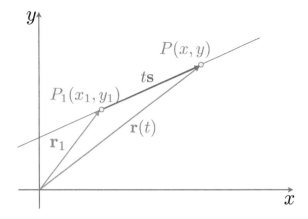

13.5.1   Unit Tangent and Normal Vectors to a Line

Figure 13.6 shows the geometry for a parametric line, where $P_1(x_1, y_1)$ and $P(x, y)$ are two points on the line, and vector $\mathbf{s}$ provides the line's direction. Let's define $\mathbf{r}_1, \mathbf{s}, \mathbf{r}(t)$:

$$\mathbf{r}_1 = x_1\mathbf{i} + y_1\mathbf{j}$$
$$\mathbf{s} = x_s\mathbf{i} + y_s\mathbf{j}$$
$$\mathbf{r}(t) = \mathbf{r}_1 + t\mathbf{s}$$
$$= (x_1 + x_s t)\mathbf{i} + (y_1 + y_s t)\mathbf{j}.$$

Differentiating $\mathbf{r}(t)$:

$$\mathbf{r}'(t) = x_s\mathbf{i} + y_s\mathbf{j}$$

whose magnitude is

$$\left\| \mathbf{r}'(t) \right\| = \sqrt{x_s^2 + y_s^2}.$$

Therefore,

$$\mathbf{T} = \frac{x_s\mathbf{i} + y_s\mathbf{j}}{\sqrt{x_s^2 + y_s^2}}.$$

Figure 13.7 shows the graph of

$$\mathbf{r}(t) = 2t\mathbf{i} + (1 + t)\mathbf{j}$$

therefore

$$\mathbf{T} = \frac{2\mathbf{i} + \mathbf{j}}{\sqrt{5}} \approx 0.8944\mathbf{i} + 0.4472\mathbf{j}$$

as shown in Fig. 13.7.

Fig. 13.7 A unit tangent and
normal vector to a line

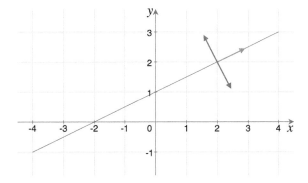

Differentiating $\mathbf{T}$ gives a zero vector, therefore our definition of $\mathbf{N}$ can't be used.
So I'll use one of the two options described above for a perpendicular vector:

$$\mathbf{N} = -0.4472\mathbf{i} + 0.8944\mathbf{j}$$

as shown in Fig. 13.7.

Using the grad operator, we can find the unit normal vector using the line equation:

$$0 = y_s(x - x_1) - x_s(y - y_1)$$
$$f(x, y) = y_s x - x_s y - y_s x_1 + x_s y_1$$
$$\nabla f = y_s \mathbf{i} - x_s \mathbf{j}$$
$$\mathbf{N} = \frac{y_s \mathbf{i} - x_s \mathbf{j}}{\sqrt{y_s^2 + x_s^2}}.$$

Evaluating $\mathbf{N}$ for $x_s = 2$, $y_s = 1$:

$$\mathbf{N} = \frac{\mathbf{i} - 2\mathbf{j}}{\sqrt{1+4}} \approx 0.4472\mathbf{i} - 0.8944\mathbf{j}$$

as shown in Fig. 13.7.

For a 3D line:

$$\mathbf{r}(t) = (x_1 + x_s t)\mathbf{i} + (y_1 + y_s t)\mathbf{j} + (z_1 + z_s t)\mathbf{k}$$
$$\mathbf{r}'(t) = x_s \mathbf{i} + y_s \mathbf{j} + z_s \mathbf{k}$$
$$\mathbf{T} = \frac{x_s \mathbf{i} + y_s \mathbf{j} + z_s \mathbf{k}}{\sqrt{x_s^2 + y_s^2 + z_s^2}}$$

however, there is no unique normal vector, only a normal plane.

13.5.2 Unit Tangent and Normal Vectors to a Parabola

We normally write a parabolic equation as

$$y = ax^2 + bx + c$$

where for different values of x there is a corresponding value of y, which describes the familiar parabolic curve. However, we require this to be described as a vector-valued function. Working in two dimensions, I will align the x-component with the $\mathbf{i}$ unit vector, and the y-component with the $\mathbf{j}$ unit vector, and use a parameter t to drive the entire process. Thus it will take the general form

$$\mathbf{r}(t) = dt\mathbf{i} + \left(at^2 + bt + c\right)\mathbf{j}$$

with suitable values for a, b, c, d. Therefore, consider the parabola

$$\mathbf{r}(t) = 2t\mathbf{i} + \left(1.5 - 1.5t^2\right)\mathbf{j}, \quad t \in [-1, 1].$$

Differentiating $\mathbf{r}(t)$:
$$\mathbf{r}'(t) = 2\mathbf{i} - 3t\mathbf{j}$$

whose magnitude is

$$\left\|\mathbf{r}'(t)\right\| = \sqrt{4 + 9t^2}.$$

Therefore,

$$\mathbf{T}(t) = \frac{2\mathbf{i} - 3t\mathbf{j}}{\sqrt{4 + 9t^2}}$$

Evaluating $\mathbf{T}(t)$ for $t = -1, 0, 1$, we get

$$\mathbf{T}(-1) = \frac{2\mathbf{i} + 3\mathbf{j}}{\sqrt{13}} \approx 0.555\mathbf{i} + 0.832\mathbf{j}$$

$$\mathbf{T}(0) = \frac{2\mathbf{i}}{\sqrt{4}} = \mathbf{i}$$

$$\mathbf{T}(1) = \frac{2\mathbf{i} - 3\mathbf{j}}{\sqrt{13}} \approx 0.555\mathbf{i} - 0.832\mathbf{j}$$

as shown in Fig. 13.8.

Computing $\mathbf{T}'(t)$ and normalising will be rather messy, so I'll choose one of two perpendicular vectors. So given the following unit tangent vectors:

Fig. 13.8 Three unit tangent
and normal vectors for a
parabola

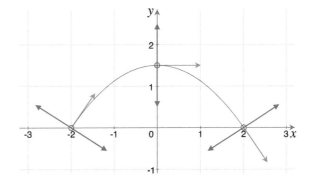

$$\mathbf{T}(-1) \approx 0.555\mathbf{i} + 0.832\mathbf{j}$$
$$\mathbf{T}(0) = \mathbf{i} + 0\mathbf{j}$$
$$\mathbf{T}(1) \approx 0.555\mathbf{i} - 0.832\mathbf{j}$$
$$\mathbf{N}(-1) \approx 0.832\mathbf{i} - 0.555\mathbf{j}$$
$$\mathbf{N}(0) = -\mathbf{j}$$
$$\mathbf{N}(1) \approx -0.832\mathbf{i} - 0.555\mathbf{j}$$

which point in the direction of the principal normal vector, as shown in Fig. 13.8.
Using the grad operator, we can find the unit normal vector as follows:

$$x = 2t$$
$$t = \tfrac{1}{2}x$$
$$y = 1.5 - 1.5t^2$$
$$= 1.5 - 1.5\frac{x^2}{4}$$
$$f(x, y) = y + 1.5\frac{x^2}{4} - 1.5$$
$$\nabla f = \tfrac{3}{4}x\mathbf{i} + \mathbf{j}$$
$$\mathbf{N}(x, y) = \frac{\tfrac{3}{4}x\mathbf{i} + \mathbf{j}}{\sqrt{\tfrac{9}{16}x^2 + 1}}.$$

Evaluating $\mathbf{N}(x, y)$ for three different positions:

$$\mathbf{N}(-2, 0) = \frac{\tfrac{-6}{4}\mathbf{i} + \mathbf{j}}{\sqrt{\tfrac{9}{16}4 + 1}} = \frac{-1.5\mathbf{i} + \mathbf{j}}{\sqrt{3.25}} \approx -0.832\mathbf{i} + 0.555\mathbf{j}$$

$$\mathbf{N}(0, 1.5) = \frac{0\mathbf{i} + \mathbf{j}}{\sqrt{1}} = \mathbf{j}$$

$$\mathbf{N}(2, 0) = \frac{\frac{6}{4}\mathbf{i} + \mathbf{j}}{\sqrt{\frac{9}{16}4 + 1}} = \frac{1.5\mathbf{i} + \mathbf{j}}{\sqrt{3.25}} \approx 0.832\mathbf{i} + 0.555\mathbf{j}$$

as shown in Fig. 13.8.

13.5.3 Unit Tangent and Normal Vectors to a Circle

Let's find the function describing the tangent vector to a circle. We start with the following definition for a circle:

$$\mathbf{r}(t) = r \cos t\mathbf{i} + r \sin t\mathbf{j}, \quad t \in [0, 2\pi].$$

Differentiating $\mathbf{r}(t)$:

$$\mathbf{r}'(t) = -r \sin t\mathbf{i} + r \cos t\mathbf{j}$$

whose magnitude is

$$||\mathbf{r}'(t)|| = \sqrt{\left(-r \sin t\right)^2 + \left(r \cos t\right)^2} = r.$$

And we see that the magnitude of the tangent vector remains constant at the circle's radius r. Therefore,

$$\mathbf{T}(t) = \frac{\mathbf{r}'(t)}{r} = -\sin t\mathbf{i} + \cos t\mathbf{j}.$$

Evaluating $\mathbf{T}(t)$ for four values of t:

$$\mathbf{T}(0°) = \mathbf{j}$$
$$\mathbf{T}(90°) = -\mathbf{i}$$
$$\mathbf{T}(180°) = -\mathbf{j}$$
$$\mathbf{T}(270°) = \mathbf{i}$$

as shown in Fig. 13.9.

To find $\mathbf{N}(t)$ we differentiate $\mathbf{T}(t) = -\sin t\mathbf{i} + \cos t\mathbf{j}$

$$\mathbf{N}(t) = \mathbf{T}'(t) = -\cos t\mathbf{i} - \sin t\mathbf{j}.$$

Evaluating $\mathbf{N}(t)$ for four values of t:

$$\mathbf{N}(0°) = -\mathbf{i}$$
$$\mathbf{N}(90°) = -\mathbf{j}$$
$$\mathbf{N}(180°) = \mathbf{i}$$
$$\mathbf{N}(270°) = \mathbf{j}$$

as shown in Fig. 13.9.

Using the grad operator, we can find the unit normal vector as follows:

$$x^2 + y^2 = r^2$$
$$f(x, y) = x^2 + y^2 - r^2$$
$$\nabla f = 2x\mathbf{i} + 2y\mathbf{j}$$
$$\mathbf{N}(x, y) = \frac{2x\mathbf{i} + 2y\mathbf{j}}{\sqrt{4x^2 + 4y^2}}.$$

Evaluating $\mathbf{N}(x, y)$ for the same positions before, where $r = 1$:

$$\mathbf{N}(1, 0) = \frac{2\mathbf{i} + 0\mathbf{j}}{\sqrt{4}} = \mathbf{i}$$
$$\mathbf{N}(0, 1) = \frac{0\mathbf{i} + 3\mathbf{j}}{\sqrt{4}} = \mathbf{j}$$
$$\mathbf{N}(-1, 0) = \frac{-2\mathbf{i} + 0\mathbf{j}}{\sqrt{4}} = -\mathbf{i}$$
$$\mathbf{N}(0, -1) = \frac{0\mathbf{i} - 2\mathbf{j}}{\sqrt{4}} = -\mathbf{j}$$

as shown in Fig. 13.9.

Fig. 13.9 Four unit tangent and normal vectors for a circle

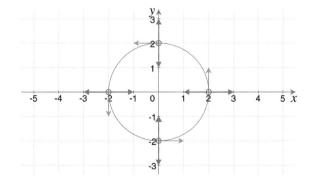

13.5.4 Unit Tangent and Normal Vectors to an Ellipse

Having found the unit tangent and normal vectors for a circle, an ellipse should be similar. Let's define an ellipse as

$$\mathbf{r}(t) = a\cos t\mathbf{i} + b\sin t\mathbf{j}, \quad t \in [0, 2\pi]$$

Differentiating $\mathbf{r}(t)$:

$$\mathbf{r}'(t) = -a\sin t\mathbf{i} + b\cos t\mathbf{j}$$

whose magnitude is

$$\begin{aligned}
||\mathbf{r}'(t)|| &= \sqrt{(-a\sin t)^2 + (b\cos t)^2} \\
&= \sqrt{a^2\sin^2 t + b^2\cos^2 t} \\
&= \sqrt{a^2(1 - \cos^2 t) + b^2\cos^2 t} \\
&= \sqrt{a^2 - (a^2 - b^2)\cos^2 t} \\
&= a\sqrt{1 - \epsilon^2\cos^2 t}
\end{aligned}$$

where $\epsilon = \sqrt{1 - b^2/a^2}$ is the eccentricity of the ellipse.
Therefore

$$\mathbf{T}(t) = \frac{-a\sin t\mathbf{i} + b\cos t\mathbf{j}}{a\sqrt{1 - \epsilon^2\cos^2 t}}.$$

As an example, let's define an ellipse with $a = 2$ and $b = 1.5$, which makes the eccentricity:

$$\epsilon = \sqrt{1 - 1.5^2/2^2} = \sqrt{0.4375}.$$

Evaluating $\mathbf{T}(t)$ for four values of t:

$$\mathbf{T}(0°) = \frac{1.5\mathbf{j}}{2\sqrt{1 - 0.4375}} = \mathbf{j}$$

$$\mathbf{T}(90°) = -\frac{2\mathbf{i}}{2\sqrt{1}} = -\mathbf{i}$$

$$\mathbf{T}(180°) = \frac{-1.5\mathbf{j}}{2\sqrt{1 - 0.4375}} = -\mathbf{j}$$

$$\mathbf{T}(270°) = \frac{2\mathbf{i}}{2\sqrt{1}} = \mathbf{i}$$

as shown in Fig. 13.10. Once again, there is no need to differentiate $\mathbf{T}(t)$ to find $\mathbf{N}(t)$. We simply use the perpendicular strategy explained above. Therefore,

$$\mathbf{N}(0°) = -\mathbf{i}$$
$$\mathbf{N}(90°) = -\mathbf{j}$$
$$\mathbf{N}(180°) = \mathbf{i}$$
$$\mathbf{N}(270°) = \mathbf{j}$$

as shown in Fig. 13.10.

Using the grad operator, we can find the unit normal vector as follows:

$$\frac{x^2}{a^2} + \frac{y^2}{b^2} = 1$$

$$f(x, y) = \frac{x^2}{a^2} + \frac{y^2}{b^2} - 1$$

$$\nabla f = \frac{2x}{a^2}\mathbf{i} + \frac{2y}{b^2}\mathbf{j}.$$

Substituting $a = 2, b = 1.5$ and $(x, y) = (2, 0), (0, 1.5), (-2, 0), (0, -1.5)$:

$$\nabla f(2, 0) = \tfrac{4}{4}\mathbf{i} + \tfrac{0}{2.25}\mathbf{j}$$
$$\mathbf{N}(2, 0) = \mathbf{i}$$
$$\nabla f(0, 1.5) = \tfrac{0}{4}\mathbf{i} + \tfrac{3}{2.25}\mathbf{j}$$
$$\mathbf{N}(0, 1.5) = \mathbf{j}$$
$$\nabla f(-2, 0) = \tfrac{-4}{4}\mathbf{i} + \tfrac{0}{2.25}\mathbf{j}$$
$$\mathbf{N}(2, 0) = -\mathbf{i}$$
$$\nabla f(0, -1.5) = \tfrac{0}{4}\mathbf{i} - \tfrac{3}{2.25}\mathbf{j}$$
$$\mathbf{N}(0, -1.5) = -\mathbf{j}$$

as shown in Fig. 13.10.

Fig. 13.10 Four unit tangent and normal vectors for an ellipse

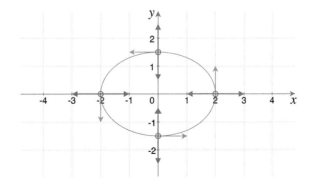

13.5.5   *Unit Tangent and Normal Vectors to a Sine Curve*

Let's calculate the tangent and normal vectors to a sine waveform. This may be of
interest in the rendering of sinusoidal waves. We define one period of a sine curve as

$$\mathbf{r}(t) = t\mathbf{i} + 2\sin t\mathbf{j}, \quad t \in [0, 2\pi].$$

Differentiating $\mathbf{r}(t)$:

$$\mathbf{r}'(t) = \mathbf{i} + 2\cos t\mathbf{j}$$

whose magnitude is

$$\|\mathbf{r}'(t)\| = \sqrt{1 + 4\cos^2 t}.$$

Therefore,

$$\mathbf{T}(t) = \frac{\mathbf{i} + 2\cos t\mathbf{j}}{\sqrt{1 + 4\cos^2 t}}.$$

Evaluating $\mathbf{T}(t)$ for four values of t:

$$\mathbf{T}(0°) = \frac{\mathbf{i} + 2\mathbf{j}}{\sqrt{5}} \approx 0.4472\mathbf{i} + 0.8944\mathbf{j}$$

$$\mathbf{T}(90°) = \frac{\mathbf{i}}{\sqrt{1}} = \mathbf{i}$$

$$\mathbf{T}(180°) = \frac{\mathbf{i} - 2\mathbf{j}}{\sqrt{5}} \approx 0.4472\mathbf{i} - 0.8944\mathbf{j}$$

$$\mathbf{T}(270°) = \frac{\mathbf{i}}{\sqrt{1}} = \mathbf{i}$$

as shown in Fig. 13.11.

Fig. 13.11 Four unit tangent
and normal vectors for a sine
curve

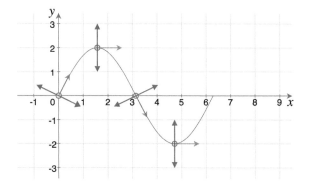

To differentiate $\mathbf{T}(t)$ and normalise it, looks as though it requires considerable work, so we'll take the easy root as before. Therefore,

$$\mathbf{N}(0°) \approx 0.8944\mathbf{i} - 0.4472\mathbf{j}$$
$$\mathbf{N}(90°) = -\mathbf{j}$$
$$\mathbf{N}(180°) \approx -0.8944\mathbf{i} - 0.4472\mathbf{j}$$
$$\mathbf{N}(270°) = \mathbf{j}$$

as shown in Fig. 13.11.

Using the grad operator, we can find the unit normal vector as follows:

$$y = 2\sin x$$
$$f(x, y) = y - 2\sin x = 0$$
$$\nabla f = -2\cos x\mathbf{i} + \mathbf{j}$$
$$\mathbf{N}(x) = \frac{-2\cos x\mathbf{i} + \mathbf{j}}{\sqrt{1 + 4\cos^2 x}}.$$

Evaluating $\mathbf{N}(x)$ for four values of x:

$$\mathbf{N}(0) = \frac{-2\mathbf{i} + \mathbf{j}}{\sqrt{5}} \approx -0.8944\mathbf{i} + 0.4472\mathbf{j}$$
$$\mathbf{N}(\pi/2) = \mathbf{j}$$
$$\mathbf{N}(\pi) \approx 0.8944\mathbf{i} + 0.4472\mathbf{j}$$
$$\mathbf{N}(3\pi/2) = -\mathbf{j}$$

as shown in Fig. 13.11.

13.5.6   Unit Tangent and Normal Vectors to a **cosh** Curve

Now let's calculate the tangent and normal vectors to a cosh curve, also called a *catenary*. We define part of a cosh curve as

$$\mathbf{r}(t) = t\mathbf{i} + 3\cosh\left(\tfrac{t}{3}\right)\mathbf{j}, \quad t \in [-3, 3]$$

Differentiating $\mathbf{r}(t)$:
$$\mathbf{r}'(t) = \mathbf{i} + \sinh\left(\tfrac{t}{3}\right)\mathbf{j}$$

whose magnitude is

Fig. 13.12 Three unit
tangent and normal vectors
for $3\cosh(x/3)$

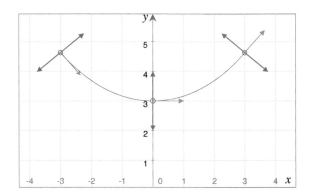

$$\|\mathbf{r}'(t)\| = \sqrt{1 + \sinh^2\left(\tfrac{t}{3}\right)} = \cosh\left(\tfrac{t}{3}\right).$$

Therefore,

$$\mathbf{T}(t) = \frac{\mathbf{i} + \sinh\left(\tfrac{t}{3}\right)\mathbf{j}}{\cosh\left(\tfrac{t}{3}\right)}.$$

Let's find $\mathbf{T}(t)$ for three values of t:

$$\mathbf{T}(-3) = \frac{\mathbf{i} - 1.1752\mathbf{j}}{1.5431} \approx 0.6481\mathbf{i} - 0.7616\mathbf{j}$$

$$\mathbf{T}(0) = \mathbf{i}$$

$$\mathbf{T}(3) = \frac{\mathbf{i} + 1.1752\mathbf{j}}{1.5431} \approx 0.6481\mathbf{i} + 0.7616\mathbf{j}$$

as shown in Fig. 13.12.

   This time, let's differentiate $\mathbf{T}(t)$ and normalise it:

$$\mathbf{T}(t) = \frac{\mathbf{i} + \sinh\left(\tfrac{t}{3}\right)\mathbf{j}}{\cosh\left(\tfrac{t}{3}\right)} = \operatorname{sech}\left(\tfrac{t}{3}\right)\mathbf{i} + \tanh\left(\tfrac{t}{3}\right)\mathbf{j}$$

$$\mathbf{T}'(t) = -\tfrac{1}{3}\left(\operatorname{sech}\left(\tfrac{t}{3}\right)\tanh\left(\tfrac{t}{3}\right)\mathbf{i} + \operatorname{sech}^2\left(\tfrac{t}{3}\right)\mathbf{j}\right)$$

$$= -\tfrac{1}{3}\left(\frac{\sinh\left(\tfrac{t}{3}\right)}{\cosh^2\left(\tfrac{t}{3}\right)}\mathbf{i} + \frac{1}{\cosh^2\left(\tfrac{t}{3}\right)}\mathbf{j}\right)$$

$$\|\mathbf{T}'(t)\| = \tfrac{1}{3}\sqrt{\frac{\sinh^2\left(\tfrac{t}{3}\right)}{\cosh^4\left(\tfrac{t}{3}\right)} + \frac{1}{\cosh^4\left(\tfrac{t}{3}\right)}} = \tfrac{1}{3}\operatorname{sech}\left(\tfrac{t}{3}\right)$$

$$\mathbf{N}(t) = -\cosh\left(\tfrac{t}{3}\right)\left(\frac{\sinh\left(\tfrac{t}{3}\right)}{\cosh^2\left(\tfrac{t}{3}\right)}\mathbf{i} + \frac{1}{\cosh^2\left(\tfrac{t}{3}\right)}\mathbf{j}\right)$$

$$= -\tanh\left(\tfrac{t}{3}\right)\mathbf{i} + \operatorname{sech}\left(\tfrac{t}{3}\right)\mathbf{j}.$$

Note that this is one of the options if we had taken the easy route!
Therefore,

$$N(-3) \approx 0.7616i + 0.6481j$$
$$N(0) = j$$
$$N(3) \approx -0.7616i + 0.6481j$$

as shown in Fig. 13.12.

Using the grad operator, we can find the unit normal vector as follows:

$$y = 3\cosh\left(\tfrac{t}{3}\right)$$
$$f(x, y) = y - 3\cosh\left(\tfrac{t}{3}\right)$$
$$\nabla f = -\sinh\left(\tfrac{t}{3}\right)i + j$$
$$N(x) = \frac{-\sinh\left(\tfrac{t}{3}\right)i + j}{\sqrt{1 + \sinh^2\left(\tfrac{t}{3}\right)}}$$
$$= \frac{-\sinh\left(\tfrac{t}{3}\right)i + j}{\cosh\left(\tfrac{t}{3}\right)}$$
$$= -\tanh\left(\tfrac{t}{3}\right)i + \operatorname{sech}\left(\tfrac{t}{3}\right)j$$

which gives the same result as before.

13.5.7   Unit Tangent and Normal Vectors to a Helix

A helix is a 3D curve and used in nature to store the genetic code of all living organisms. It can have a variable radius, and also a variable pitch. However, a fixed radius and constant-pitch helix is a popular curve used for illustrating tangent and normal vectors. Let's define a helix as

$$r(t) = 2\cos t\, i + 2\sin t\, j + t k, \quad t \in [0, 4\pi].$$

Differentiating $r(t)$:
$$r'(t) = -2\sin t\, i + 2\cos t\, j + k$$

whose magnitude is

$$\|r(t)\| = \sqrt{4\sin^2 t + 4\cos^2 t + 1} = \sqrt{5}.$$

Therefore,

$$T(t) = \tfrac{1}{\sqrt{5}}\left(-2\sin t\, i + 2\cos t\, j + k\right)$$

Fig. 13.13 Three unit
tangent and normal vectors
for a helix

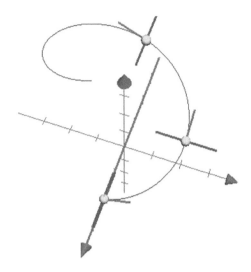

Evaluating $\mathbf{T}(t)$ for different values of t:

$$\mathbf{T}(0) = \tfrac{1}{\sqrt{5}}(2\mathbf{j} + \mathbf{k}) \approx 0.8944\mathbf{j} + 0.4472\mathbf{k}$$
$$\mathbf{T}(\pi/2) = \tfrac{1}{\sqrt{5}}(-2\mathbf{i} + \mathbf{k}) \approx -0.8944\mathbf{i} + 0.4472\mathbf{k}$$
$$\mathbf{T}(\pi) = \tfrac{1}{\sqrt{5}}(-2\mathbf{j} + \mathbf{k}) \approx -0.8944\mathbf{j} + 0.4472\mathbf{k}$$

as shown in Fig. 13.13.

Differentiating $\mathbf{T}(t)$ and normalising:

$$\mathbf{T}'(t) = \tfrac{1}{\sqrt{5}}(-2\cos t\mathbf{i} - 2\sin t\mathbf{j})$$
$$\|\mathbf{T}'(t)\| = \tfrac{1}{\sqrt{5}}\sqrt{4\cos^2 t + 4\sin^2 t} = \tfrac{2}{\sqrt{5}}$$
$$\mathbf{N}(t) = \frac{\tfrac{1}{\sqrt{5}}(-2\cos t\mathbf{i} - 2\sin t\mathbf{j})}{\tfrac{2}{\sqrt{5}}} = -\cos t\mathbf{i} - \sin t\mathbf{j}.$$

Evaluating $\mathbf{N}(t)$ for different values of t:

$$\mathbf{N}(0) = -\mathbf{i}$$
$$\mathbf{N}(\pi/2) = -\mathbf{j}$$
$$\mathbf{N}(t) = \mathbf{i}$$

as shown in Fig. 13.13.

13.5.8 Unit Tangent and Normal Vectors to a Quadratic Bézier Curve

Quadratic curves are normally expressed using a basis function $\mathbf{B}(t)$, which generates values using the parameter t. I will derive the derivative of a general basis function in Chap. 14. A 2D quadratic curve is expressed using a column vector as

$$\mathbf{r}(t) = \begin{bmatrix} x(t) \\ y(t) \end{bmatrix}, \quad t \in [0, \ 1]$$

$$x(t) = \mathbf{B}_{2,0}(t)x_0 + \mathbf{B}_{2,1}(t)x_1 + \mathbf{B}_{2,2}(t)x_2$$

$$y(t) = \mathbf{B}_{2,0}(t)y_0 + \mathbf{B}_{2,1}(t)y_1 + \mathbf{B}_{2,2}(t)y_2$$

$$\mathbf{B}_{2,0} = (1-t)^2$$

$$\mathbf{B}_{2,1} = 2t(1-t)$$

$$\mathbf{B}_{2,2} = t^2$$

algebraically:

$$\mathbf{r}(t) = \mathbf{B}_{2,0}(t)\mathbf{P}_0 + \mathbf{B}_{2,1}(t)\mathbf{P}_1 + \mathbf{B}_{2,2}(t)\mathbf{P}_2, \quad t \in [0, 1]$$

where $\mathbf{P}_0, \mathbf{P}_1, \mathbf{P}_2$ are position vectors for the control point P_0, P_1, P_2.

But a Cartesian vector is rarely used. So, for the time being, I will use algebraic notation.

Let's start with the following 2D quadratic Bézier curve:

$$\mathbf{r}(t) = \mathbf{P}_0(1-t)^2 + 2\mathbf{P}_1 t(1-t) + \mathbf{P}_2 t^2, \quad t \in [0, 1].$$

Differentiating $\mathbf{r}(t)$:

$$\begin{aligned}
\mathbf{r}'(t) &= -2\mathbf{P}_0(1-t) + 2\mathbf{P}_1(1-2t) + 2\mathbf{P}_2 t \\
&= -2\mathbf{P}_0 + 2\mathbf{P}_0 t + 2\mathbf{P}_1 - 4\mathbf{P}_1 t + 2\mathbf{P}_2 t \\
&= 2(\mathbf{P}_1 - \mathbf{P}_0)(1-t) + 2(\mathbf{P}_2 - \mathbf{P}_1)t \\
&= 2\big((\mathbf{P}_1 - \mathbf{P}_0)(1-t) + (\mathbf{P}_2 - \mathbf{P}_1)t\big) \\
x'(t) &= 2\big((x_1 - x_0)(1-t) + (x_2 - x_1)t\big) \\
y'(t) &= 2\big((y_1 - y_0)(1-t) + (y_2 - y_1)t\big)
\end{aligned}$$

whose magnitude is

$$\|\mathbf{r}'(t)\| = \sqrt{\big(x'(t)\big)^2 + \big(y'(t)\big)^2}.$$

Therefore,

$$\mathbf{T}(t) = \frac{2[(\mathbf{P}_1 - \mathbf{P}_0)(1 - t) + (\mathbf{P}_2 - \mathbf{P}_1)t]}{\sqrt{(x'(t))^2 + (y'(t))^2}}.$$

Now let's substitute specific values for $\mathbf{P}_0, \mathbf{P}_1, \mathbf{P}_2$, $P_0 = (0, 0)$, $P_1 = (1, 1)$, $P_2 = (2, 1)$. Therefore,

$$x'(t) = 2\big((1 - 0)(1 - t) + (2 - 1)t\big) = 2$$
$$y'(t) = 2\big((1 - 0)(1 - t) + (1 - 1)t\big) = 2(1 - t)$$
$$\mathbf{T}(t) = \frac{2\mathbf{i} + 2(1 - t)\mathbf{j}}{\sqrt{4 + 4(1 - t)^2}} = \frac{\mathbf{i} + (1 - t)\mathbf{j}}{\sqrt{1 + (1 - t)^2}}$$

Evaluating $\mathbf{T}(t)$ for different values of t:

$$\mathbf{T}(0) = \frac{\mathbf{i} + \mathbf{j}}{\sqrt{2}} \approx 0.7071\mathbf{i} + 0.7071\mathbf{j}$$
$$\mathbf{T}(0.5) = \frac{\mathbf{i} + 0.5\mathbf{j}}{\sqrt{1.25}} \approx 0.8944\mathbf{i} + 0.4472\mathbf{j}$$
$$\mathbf{T}(1) = \mathbf{i}$$

as shown in Fig. 13.14. Differentiating $\mathbf{T}(t)$ and normalising looks like a lot of work, so we'll take the easy route. Therefore,

$$\mathbf{N}(0) \approx 0.7071\mathbf{i} - 0.7071\mathbf{j}$$
$$\mathbf{N}(0.5) \approx 0.4472\mathbf{i} - 0.8944\mathbf{j}$$
$$\mathbf{N}(1) = -\mathbf{j}$$

as shown in Fig. 13.14.

Fig. 13.14 Three unit tangent and normal vectors for a quadratic Bézier curve

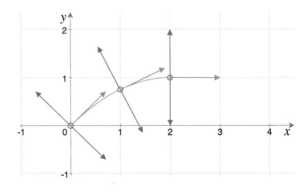

13.6   Unit Tangent and Normal Vectors to a Surface

In the following examples I show how to calculate the tangent and normal vectors to a bilinear patch, a Bézier patch, a sphere and a torus. They each require a slightly different approach, which is explained for each surface.

13.6.1   Unit Normal Vectors to a Bilinear Patch

Bilinear patches are constructed from a pair of lines using linear interpolation. For example, given two lines defined by their position vectors:

$$L_1 = (\mathbf{P}_0, \mathbf{P}_1), \qquad L_2 = (\mathbf{P}_2, \mathbf{P}_3)$$

we can linearly interpolate along the lines using

$$\mathbf{A}(u) = (1 - u)\mathbf{P}_0 + u\mathbf{P}_1, \quad u \in [0, 1]$$
$$\mathbf{B}(u) = (1 - u)\mathbf{P}_2 + u\mathbf{P}_3, \quad u \in [0, 1]$$

and then linearly interpolate between $\mathbf{A}(u)$ and $\mathbf{B}(u)$:

$$\begin{aligned}
\mathbf{r}(u, v) &= (1 - v)\mathbf{A}(u) + v\mathbf{B}(u), \quad v \in [0, 1] \\
&= (1 - v)\big((1 - u)\mathbf{P}_0 + u\mathbf{P}_1\big) + v\big((1 - u)\mathbf{P}_2 + u\mathbf{P}_3\big) \\
&= (1 - v)(1 - u)\mathbf{P}_0 + u(1 - v)\mathbf{P}_1 + v(1 - u)\mathbf{P}_2 + uv\mathbf{P}_3.
\end{aligned}$$

We now compute the partial derivatives for u and v:

$$\begin{aligned}
\frac{\partial \mathbf{r}}{\partial u} &= -(1 - v)\mathbf{P}_0 + (1 - v)\,\mathbf{P}_1 - v\mathbf{P}_2 + v\mathbf{P}_3 \\
&= (1 - v)(\mathbf{P}_1 - \mathbf{P}_0) + v(\mathbf{P}_3 - \mathbf{P}_2) \\
\frac{\partial \mathbf{r}}{\partial v} &= -(1 - u)\mathbf{P}_0 - u\mathbf{P}_1 + (1 - u)\mathbf{P}_2 + u\mathbf{P}_3 \\
&= (1 - u)(\mathbf{P}_2 - \mathbf{P}_0) + u(\mathbf{P}_3 - \mathbf{P}_1).
\end{aligned}$$

$\frac{\partial \mathbf{r}}{\partial u}$ and $\frac{\partial \mathbf{r}}{\partial v}$ encode a pair of orthogonal tangent vectors, whose cross-product is a vector normal.

Let's demonstrate this with an example. Given:

$$\begin{aligned}
\mathbf{P}_0 &= 0\mathbf{i} + 0\mathbf{j} + 0\mathbf{k} \\
\mathbf{P}_1 &= 0\mathbf{i} + 2\mathbf{j} + \mathbf{k} \\
\mathbf{P}_2 &= 2\mathbf{i} + 0\mathbf{j} + 0\mathbf{k} \\
\mathbf{P}_3 &= 2\mathbf{i} + 2\mathbf{j} - \mathbf{k}
\end{aligned}$$

then

$$\frac{\partial \mathbf{r}}{\partial u} = 0\mathbf{i} + \big(2(1 - v) + 2v\big)\mathbf{j} + \big((1 - v) - v\big)\mathbf{k}$$

$$= 0\mathbf{i} + 2\mathbf{j} + (1 - 2v)\mathbf{k}$$

$$\frac{\partial \mathbf{r}}{\partial v} = \big(2(1 - u) + 2u\big)\mathbf{i} + 0\mathbf{j} - 2u\mathbf{k}$$

$$= 2\mathbf{i} + 0\mathbf{j} - 2u\mathbf{k}.$$

We can now calculate their cross product:

$$\mathbf{T}'(u, v) = \begin{vmatrix} \mathbf{i} & \mathbf{j} & \mathbf{k} \\ 2 & 0 & -2u \\ 0 & 2 & 1 - 2v \end{vmatrix} = 4u\mathbf{i} + (4v - 2)\mathbf{j} + 4\mathbf{k}$$

which is a vector orthogonal to the tangent vectors, depending on the value of u and v. Let's calculate the unit normal vector by dividing the normal vector by its magnitude, for different values of u and v.

$$\mathbf{N}(0, 0) = \frac{-2\mathbf{j} + 4\mathbf{k}}{\sqrt{20}} \approx -0.4472\mathbf{j} + 0.8944\mathbf{k}$$

$$\mathbf{N}(1, 0) = \frac{4\mathbf{i} - 2\mathbf{j} + 4\mathbf{k}}{\sqrt{34}} \approx 0.686\mathbf{i} - 0.343\mathbf{j} + 0.686\mathbf{k}$$

$$\mathbf{N}(0, 1) = \frac{2\mathbf{j} + 4\mathbf{k}}{\sqrt{20}} \approx 0.4472\mathbf{j} + 0.8944\mathbf{k}$$

$$\mathbf{N}(1, 1) = \frac{4\mathbf{i} + 2\mathbf{j} + 4\mathbf{k}}{\sqrt{34}} \approx 0.686\mathbf{i} + 0.343\mathbf{j} + 0.686\mathbf{k}$$

$$\mathbf{N}(0.5, 0.5) = \frac{2\mathbf{i} + 4\mathbf{k}}{\sqrt{2}} \approx 0.4472\mathbf{i} + 0.8944\mathbf{k}$$

as shown in Fig. 13.15.

13.6.2 Unit Normal Vectors to a Quadratic Bézier Patch

Bézier proposed a matrix of nine control points to determine the geometry of a quadratic patch, as shown in Fig. 13.16. Any point on the patch is defined by

$$\mathbf{P}_{uv} = [u^2 \quad u \quad 1] \begin{bmatrix} 1 & -2 & 1 \\ -2 & 2 & 0 \\ 1 & 0 & 0 \end{bmatrix} \begin{bmatrix} P_{00} & P_{01} & P_{02} \\ P_{10} & P_{11} & P_{12} \\ P_{20} & P_{21} & P_{22} \end{bmatrix} \begin{bmatrix} 1 & -2 & 1 \\ -2 & 2 & 0 \\ 1 & 0 & 0 \end{bmatrix} \begin{bmatrix} v^2 \\ v \\ 1 \end{bmatrix}.$$

Fig. 13.15 Five unit normal
vectors for a bilinear surface

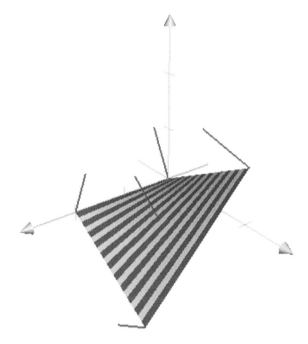

Fig. 13.16 A quadratic
Bézier surface patch

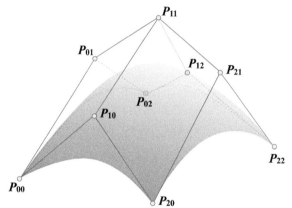

The individual x-, y- and z-coordinates are obtained by substituting the x-, y- and
z-values for the central **P** matrix.

Let's illustrate the process with an example. Given the following points:

$$P_{00} = (0,\ 0,\ 0),\quad P_{01} = (1,\ 1,\ 0),\quad P_{02} = (2,\ 0,\ 0)$$
$$P_{10} = (0,\ 1,\ 1),\quad P_{11} = (1,\ 2,\ 1),\quad P_{12} = (2,\ 1,\ 1)$$
$$P_{20} = (0,\ 0,\ 2),\quad P_{21} = (1,\ 1,\ 2),\quad P_{22} = (2,\ 0,\ 2)$$

we can write

$$x_{uv} = [u^2 \ u \ 1] \begin{bmatrix} 1 & -2 & 1 \\ -2 & 2 & 0 \\ 1 & 0 & 0 \end{bmatrix} \begin{bmatrix} 0 & 1 & 2 \\ 0 & 1 & 2 \\ 0 & 1 & 2 \end{bmatrix} \begin{bmatrix} 1 & -2 & 1 \\ -2 & 2 & 0 \\ 1 & 0 & 0 \end{bmatrix} \begin{bmatrix} v^2 \\ v \\ 1 \end{bmatrix}$$

$$x_{uv} = [u^2 \ u \ 1] \begin{bmatrix} 0 & 0 & 0 \\ 0 & 0 & 0 \\ 0 & 2 & 0 \end{bmatrix} \begin{bmatrix} v^2 \\ v \\ 1 \end{bmatrix}$$

$$x_{uv} = 2v$$

$$y_{uv} = [u^2 \ u \ 1] \begin{bmatrix} 1 & -2 & 1 \\ -2 & 2 & 0 \\ 1 & 0 & 0 \end{bmatrix} \begin{bmatrix} 0 & 1 & 0 \\ 1 & 2 & 1 \\ 0 & 1 & 0 \end{bmatrix} \begin{bmatrix} 1 & -2 & 1 \\ -2 & 2 & 0 \\ 1 & 0 & 0 \end{bmatrix} \begin{bmatrix} v^2 \\ v \\ 1 \end{bmatrix}$$

$$y_{uv} = [u^2 \ u \ 1] \begin{bmatrix} 0 & 0 & -2 \\ 0 & 0 & 2 \\ -2 & 2 & 0 \end{bmatrix} \begin{bmatrix} v^2 \\ v \\ 1 \end{bmatrix}$$

$$y_{uv} = 2(u + v - u^2 - v^2)$$

$$z_{uv} = [u^2 \ u \ 1] \begin{bmatrix} 1 & -2 & 1 \\ -2 & 2 & 0 \\ 1 & 0 & 0 \end{bmatrix} \begin{bmatrix} 0 & 0 & 0 \\ 1 & 1 & 1 \\ 2 & 2 & 2 \end{bmatrix} \begin{bmatrix} 1 & -2 & 1 \\ -2 & 2 & 0 \\ 1 & 0 & 0 \end{bmatrix} \begin{bmatrix} v^2 \\ v \\ 1 \end{bmatrix}$$

$$z_{uv} = [u^2 \ u \ 1] \begin{bmatrix} 0 & 0 & 0 \\ 0 & 0 & 2 \\ 0 & 0 & 0 \end{bmatrix} \begin{bmatrix} v^2 \\ v \\ 1 \end{bmatrix}$$

$$z_{uv} = 2u.$$

Therefore, any point on the surface patch has coordinates

$$\mathbf{p}_{uv} = 2v\mathbf{i} + 2\left(u + v - u^2 - v^2\right)\mathbf{j} + 2u\mathbf{k}.$$

To calculate a unit vector normal to the surface we first calculate two tangent vectors using $\frac{\partial \mathbf{p}}{\partial u}$ and $\frac{\partial \mathbf{p}}{\partial v}$, take their cross product, and normalise the resulting vector.

$$\frac{\partial \mathbf{p}}{\partial u} = 0\mathbf{i} + 2(1 - 2u)\mathbf{j} + 2\mathbf{k}$$

$$\frac{\partial \mathbf{p}}{\partial v} = 2\mathbf{i} + 2(1 - 2v)\mathbf{j} + 0\mathbf{k}.$$

We can now compute their cross product:

$$\begin{vmatrix} \mathbf{i} & \mathbf{j} & \mathbf{k} \\ 0 & 2 - 4u & 2 \\ 2 & 2 - 4v & 0 \end{vmatrix} = (8v - 4)\mathbf{i} + 4\mathbf{j} + (8u - 4)\mathbf{k}$$

Fig. 13.17 Five unit normal
vectors for a quadratic
Bézier patch

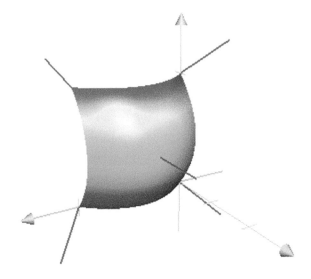

which is a vector orthogonal to the tangent vectors, depending on the value of u
and v. Let's calculate the unit normal vector by dividing the normal vector by its
magnitude, for different values of u and v.

$$\mathbf{N}(0, 0) = \frac{-4\mathbf{i} + 4\mathbf{j} - 4\mathbf{k}}{\sqrt{48}} \approx -0.5774\mathbf{i} + 0.5774\mathbf{j} - 0.5774\mathbf{k}$$

$$\mathbf{N}(1, 0) = \frac{-4\mathbf{i} + 4\mathbf{j} + 4\mathbf{k}}{\sqrt{48}} \approx -0.5774\mathbf{i} + 0.5774\mathbf{j} + 0.5774\mathbf{k}$$

$$\mathbf{N}(0, 1) = \frac{4\mathbf{i} + 4\mathbf{j} - 4\mathbf{k}}{\sqrt{48}} \approx 0.5774\mathbf{i} + 0.5774\mathbf{j} - 0.5774\mathbf{k}$$

$$\mathbf{N}(1, 1) = \frac{4\mathbf{i} + 4\mathbf{j} + 4\mathbf{k}}{\sqrt{48}} \approx 0.5774\mathbf{i} + 0.5774\mathbf{j} + 0.5774\mathbf{k}$$

$$\mathbf{N}(0.5, 0.5) = \frac{0\mathbf{i} + 4\mathbf{j} + 0\mathbf{k}}{\sqrt{16}} \approx 0\mathbf{i} + \mathbf{j} + 0\mathbf{k}$$

as shown in Fig. 13.17.

13.6.3 Unit Tangent and Normal Vector to a Sphere

It should not be too difficult to find the tangent and normal vectors for a sphere. So
let's start with the equation for a sphere with radius r in Cartesian coordinates as

$$x^2 + y^2 + z^2 = r^2.$$

Therefore, we can declare a function:

$$f(x, y, z) = x^2 + y^2 + z^2 - r^2.$$

This is best solved using the notation of a gradient vector ∇, where

$$\nabla f = \frac{\partial f}{\partial x}\mathbf{i} + \frac{\partial f}{\partial y}\mathbf{j} + \frac{\partial f}{\partial z}\mathbf{k}.$$

Therefore,

$$\frac{\partial f}{\partial x} = 2x$$
$$\frac{\partial f}{\partial y} = 2y$$
$$\frac{\partial f}{\partial z} = 2z$$
$$\nabla f = 2x\mathbf{i} + 2y\mathbf{j} + 2z\mathbf{k}$$

which is a vector normal to the sphere.

Let's compute the unit normal vector for different points on a sphere using

$$\mathbf{N} = \frac{\nabla f}{\|\nabla f\|}.$$

But in order to identify points on the sphere's surface, it is easier to use spherical coordinates, where

$$x = r \sin \phi \cdot \cos \theta$$
$$y = r \sin \phi \cdot \sin \theta$$
$$z = r \cos \phi$$

therefore,

$$\nabla f(r, \phi, \theta) = 2r \sin \phi \cdot \cos \theta \mathbf{i} + 2r \sin \phi \cdot \sin \theta \mathbf{j} + 2r \cos \phi \mathbf{k}$$

$$\mathbf{N}\left(5, \tfrac{\pi}{2}, 0\right) = \frac{10\mathbf{i} + 0\mathbf{j} + 0\mathbf{k}}{\sqrt{100}} = \mathbf{i}$$

$$\mathbf{N}\left(5, \tfrac{\pi}{2}, \tfrac{\pi}{2}\right) = \frac{0\mathbf{i} + 10\mathbf{j} + 0\mathbf{k}}{\sqrt{100}} = \mathbf{j}$$

$$\mathbf{N}\left(5, 0, 0\right) = \frac{0\mathbf{i} + 0\mathbf{j} + 10\mathbf{k}}{\sqrt{100}} = \mathbf{k}$$

Fig. 13.18 Five unit normal
vectors for a sphere

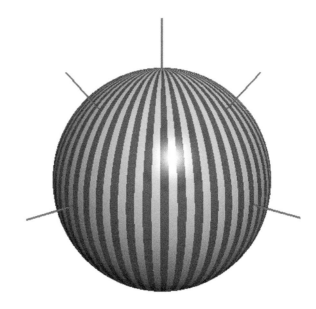

$$\mathbf{N}\left(5, \tfrac{\pi}{4}, 0\right) = \frac{5\sqrt{2}\mathbf{i} + 0\mathbf{j} + 5\sqrt{2}\mathbf{k}}{\sqrt{100}} \approx 0.7071\mathbf{i} + 0\mathbf{j} + 0.071\mathbf{k}$$

$$\mathbf{N}\left(5, \tfrac{\pi}{4}, \tfrac{\pi}{2}\right) = \frac{0\mathbf{i} + 5\sqrt{2}\mathbf{j} + 5\sqrt{2}\mathbf{k}}{\sqrt{100}} \approx 0\mathbf{i} + 0.7071\mathbf{j} + 0.071\mathbf{k}$$

as shown in Fig. 13.18. There is no unique tangent vector, only a unique tangent
plane.

13.6.4   Unit Tangent and Normal Vectors to a Torus

Lastly, let's find the tangent and normal vectors for a torus. The equation for a torus
with major radius R and minor radius r is

$$\mathbf{r}(\theta, \phi) = \begin{bmatrix} (R + r\cos\theta) \cdot \cos\phi \\ (R + r\cos\theta) \cdot \sin\phi \\ r\sin\theta \end{bmatrix}, \quad (\theta, \phi) \in [0, 2\pi].$$

The tangent vectors are given by $\frac{\partial \mathbf{r}}{\partial \phi}$ and $\frac{\partial \mathbf{r}}{\partial \theta}$:

$$\frac{\partial \mathbf{r}}{\partial \phi} = \begin{bmatrix} -(R + r\cos\theta) \cdot \sin\phi \\ (R + r\cos\theta) \cdot \cos\phi \\ 0 \end{bmatrix}, \quad \frac{\partial \mathbf{r}}{\partial \theta} = \begin{bmatrix} -r\sin\theta \cdot \cos\phi \\ -r\sin\theta \cdot \sin\phi \\ r\cos\theta \end{bmatrix}.$$

Fig. 13.19 The tangent
vectors (green) and normal
vector (red) on the left are
for $\theta = \phi = 0$. The vectors
on the right are for
$\theta = \phi = \pi/2$

For example, let $R = 3$ and $r = 1$, then

$$\frac{\partial \mathbf{r}}{\partial \phi} = \begin{bmatrix} -(3 + \cos \theta) \cdot \sin \phi \\ (3 + \cos \theta) \cdot \cos \phi \\ 0 \end{bmatrix}, \qquad \frac{\partial \mathbf{r}}{\partial \theta} = \begin{bmatrix} -\sin \theta \cdot \cos \phi \\ -\sin \theta \cdot \sin \phi \\ \cos \theta \end{bmatrix}$$

and when $\theta = \phi = 0$:

$$\frac{\partial \mathbf{r}}{\partial \phi} = \begin{bmatrix} 0 \\ 4 \\ 0 \end{bmatrix}, \qquad \frac{\partial \mathbf{r}}{\partial \theta} = \begin{bmatrix} 0 \\ 0 \\ 1 \end{bmatrix}$$

which are shown in Fig. 13.19 as unit vectors.

If we now compute the cross product $\frac{\partial \mathbf{r}}{\partial \phi} \times \frac{\partial \mathbf{r}}{\partial \theta}$, we obtain the normal vector at that point:

$$\mathbf{N} = \begin{vmatrix} \mathbf{i} & \mathbf{j} & \mathbf{k} \\ 0 & 4 & 0 \\ 0 & 0 & 1 \end{vmatrix} = 4\mathbf{i}$$

which is shown in Fig. 13.19.

Let's compute a similar set of vectors for $\theta = \phi = \pi/2$:

$$\frac{\partial \mathbf{r}}{\partial \phi} = \begin{bmatrix} -\left(3 + \cos\left(\frac{\pi}{2}\right)\right) \cdot \sin\left(\frac{\pi}{2}\right) \\ \left(3 + \cos\left(\frac{\pi}{2}\right)\right) \cdot \cos\left(\frac{\pi}{2}\right) \\ 0 \end{bmatrix}, \qquad \frac{\partial \mathbf{r}}{\partial \theta} = \begin{bmatrix} -\sin\left(\frac{\pi}{2}\right) \cdot \cos\left(\frac{\pi}{2}\right) \\ -\sin\left(\frac{\pi}{2}\right) \cdot \sin\left(\frac{\pi}{2}\right) \\ \cos \theta \end{bmatrix}$$

$$\frac{\partial \mathbf{r}}{\partial \phi} = \begin{bmatrix} -3 \\ 0 \\ 0 \end{bmatrix}, \qquad \frac{\partial \mathbf{r}}{\partial \theta} = \begin{bmatrix} 0 \\ -1 \\ 0 \end{bmatrix}$$

which are shown in Fig. 13.19 as unit vectors.

If we now compute the cross product $\frac{\partial \mathbf{r}}{\partial \phi} \times \frac{\partial \mathbf{r}}{\partial \theta}$, we obtain the normal vector at that point:

$$\mathbf{N} = \begin{vmatrix} \mathbf{i} & \mathbf{j} & \mathbf{k} \\ -3 & 0 & 0 \\ 0 & -1 & 0 \end{vmatrix} = 3\mathbf{k}$$

which is shown in Fig. 13.19.

13.7 Summary

This chapter has shown how to calculate tangent and normal vectors to various curves and surfaces. The very same techniques can be applied to other curves and surfaces, but there is no guarantee that normalising vectors will always be an easy calculation.

13.7.1 Summary of Formulae

Unit Tangent Vector

$$\mathbf{r}(t) = x(t)\mathbf{i} + y(t)\mathbf{j} + z(t)\mathbf{k}$$
$$\mathbf{T}(t) = \frac{\mathbf{r}'(t)}{||\mathbf{r}'(t)||}.$$

Unit Normal Vector

$$\mathbf{N}(t) = \frac{\mathbf{T}'(t)}{||\mathbf{T}'(t)||}.$$

Unit Tangent and Normal Vector to a Line

2D line:

$$\mathbf{r}(t) = (x_1 + x_s t)\mathbf{i} + (y_1 + y_s t)\mathbf{j}$$
$$\mathbf{r}'(t) = x_s\mathbf{i} + y_s\mathbf{j}$$
$$\mathbf{T} = \frac{x_s\mathbf{i} + y_s\mathbf{j}}{\sqrt{x_s^2 + y_s^2}} = \lambda_1\mathbf{i} + \lambda_2\mathbf{j}$$
$$\mathbf{N} = -\lambda_2\mathbf{i} + \lambda_1\mathbf{j}, \quad \text{or} \quad = \lambda_2\mathbf{i} - \lambda_1\mathbf{j}.$$

3D line:

$$\mathbf{r}(t) = (x_1 + x_s t)\mathbf{i} + (y_1 + y_s t)\mathbf{j} + (z_1 + z_s t)\mathbf{k}$$
$$\mathbf{r}'(t) = x_s\mathbf{i} + y_s\mathbf{j} + z_s\mathbf{k}$$
$$\mathbf{T} = \frac{x_s\mathbf{i} + y_s\mathbf{j} + z_s\mathbf{k}}{\sqrt{x_s^2 + y_s^2 + z_s^2}}.$$

Unit Tangent and Normal Vector to a Circle

$$\mathbf{r}(t) = r \cos t\mathbf{i} + r \sin t\mathbf{j}$$
$$\mathbf{T}(t) = -\sin t\mathbf{i} + \cos t\mathbf{j}$$
$$\mathbf{N}(t) = -\cos t\mathbf{i} - \sin t\mathbf{j}.$$

Unit Tangent and Normal Vector to an Ellipse

$$\mathbf{r}(t) = a \cos t\mathbf{i} + b \sin t\mathbf{j}$$
$$\epsilon = \sqrt{1 - b^2/a^2}$$
$$\mathbf{T}(t) = \frac{-a \sin t\mathbf{i} + b \cos t\mathbf{j}}{a\sqrt{1 - \epsilon^2 \cos^2 t}} = \lambda_1\mathbf{i} + \lambda_2\mathbf{j}$$
$$\mathbf{N} = -\lambda_2\mathbf{i} + \lambda_1\mathbf{j} \quad \text{or} \quad = \lambda_2\mathbf{i} - \lambda_1\mathbf{j}.$$

Unit Tangent and Normal Vector to a Quadratic Bézier Curve

$$\mathbf{r}(t) = \mathbf{P}_0(1 - t)^2 + 2\mathbf{P}_1 t(1 - t) + \mathbf{P}_2 t^2, \quad t \in [0, 1]$$
$$x(t) = \mathbf{B}_{2,0}(t)x_0 + \mathbf{B}_{2,1}(t)x_1 + \mathbf{B}_{2,2}(t)x_2$$
$$y(t) = \mathbf{B}_{2,0}(t)y_0 + \mathbf{B}_{2,1}(t)y_1 + \mathbf{B}_{2,2}(t)y_2$$
$$x'(t) = 2[(x_1 - x_0)(1 - t) + (x_2 - x_1)t]$$
$$y'(t) = 2[(y_1 - y_0)(1 - t) + (y_2 - y_1)t]$$
$$\mathbf{T}(t) = \frac{2\big((\mathbf{P}_1 - \mathbf{P}_0)(1 - t) + (\mathbf{P}_2 - \mathbf{P}_1)t\big)}{\sqrt{\big(x'(t)\big)^2 + \big(y'(t)\big)^2}} = \lambda_1\mathbf{i} + \lambda_2\mathbf{j}$$
$$\mathbf{N} = -\lambda_2\mathbf{i} + \lambda_1\mathbf{j}, \quad \text{or} \quad = \lambda_2\mathbf{i} - \lambda_1\mathbf{j}.$$

Unit Tangent and Normal Vector to a Bilinear Patch

$$L_1 = (\mathbf{P}_0, \mathbf{P}_1)$$
$$L_2 = (\mathbf{P}_2, \mathbf{P}_3)$$
$$\mathbf{r}(u, v) = (1 - v)(1 - u)\mathbf{P}_0 + u(1 - v)\mathbf{P}_1 + v(1 - u)\mathbf{P}_2 + uv\mathbf{P}_3$$
$$\frac{\partial \mathbf{r}}{\partial u} = (1 - v)(\mathbf{P}_1 - \mathbf{P}_0) + v(\mathbf{P}_3 - \mathbf{P}_2) = \lambda_{u1}\mathbf{i} + \lambda_{u2}\mathbf{j} + \lambda_{u3}\mathbf{k}$$
$$\frac{\partial \mathbf{r}}{\partial v} = (1 - u)(\mathbf{P}_2 - \mathbf{P}_0) + u(\mathbf{P}_3 - \mathbf{P}_1) = \lambda_{v1}\mathbf{i} + \lambda_{v2}\mathbf{j} + \lambda_{v3}\mathbf{k}$$
$$\mathbf{T}'(u, v) = \frac{\partial \mathbf{r}}{\partial u} \times \frac{\partial \mathbf{r}}{\partial v} = \begin{vmatrix} \mathbf{i} & \mathbf{j} & \mathbf{k} \\ \lambda_{u1} & \lambda_{u2} & \lambda_{u3} \\ \lambda_{v1} & \lambda_{v2} & \lambda_{v3} \end{vmatrix}$$
$$\mathbf{N}(t) = \frac{\mathbf{T}'(t)}{\|\mathbf{T}'(t)\|}.$$

Unit Normal Vector to a Sphere

$$f(x, y, z) = x^2 + y^2 + z^2 - r^2$$
$$\nabla f = 2x\mathbf{i} + 2y\mathbf{j} + 2z\mathbf{k}$$
$$\mathbf{N}(x, y, z) = \frac{\nabla f}{\|\nabla f\|}$$
$$x = r \sin\phi \cdot \cos\theta$$
$$y = r \sin\phi \cdot \sin\theta$$
$$z = r \cos\phi$$
$$\nabla f(r, \phi, \theta) = 2r \sin\phi \cdot \cos\theta\mathbf{i} + 2r \sin\phi \cdot \sin\theta\mathbf{j} + 2r \cos\phi\mathbf{k}.$$

Unit Tangent and Normal Vector to a Torus

$$\mathbf{r}(\theta, \phi) = \begin{bmatrix} (R + r \cos\theta) \cdot \cos\phi \\ (R + r \cos\theta) \cdot \sin\phi \\ r \sin\theta \end{bmatrix}, \quad (\theta, \phi) \in [0, 2\pi]$$

$$\frac{\partial \mathbf{r}}{\partial \phi} = \begin{bmatrix} -(R + r \cos\theta) \cdot \sin\phi \\ (R + r \cos\theta) \cdot \cos\phi \\ 0 \end{bmatrix}, \quad \frac{\partial \mathbf{r}}{\partial \theta} = \begin{bmatrix} -r \sin\theta \cdot \cos\phi \\ -r \sin\theta \cdot \sin\phi \\ r \cos\theta \end{bmatrix}$$

$$\mathbf{N}(\theta, \phi) = \frac{\partial \mathbf{r}}{\partial \phi} \times \frac{\partial \mathbf{r}}{\partial \theta}.$$

Chapter 14
Continuity

14.1 Introduction

In this chapter I explain how geometric continuity is ensured between segments of B-splines and Bézier curves. To begin the analysis, we return to the definition of uniform B-splines and how polynomials are chosen to provide the geometric continuity between curve segments.

14.2 B-Splines

B-splines, like Bézier curves, use polynomials to generate a curve segment. But, unlike Bézier curves, B-splines employ a series of control points that determine the curve's local geometry. This feature ensures that only a small portion of the curve is changed when a control point is moved.

There are two types of B-splines: *rational* and *non-rational* splines, which divide into two further categories: *uniform* and *non-uniform*. Rational B-splines are formed from the ratio of two polynomials such as

$$x(t) = \frac{X(t)}{W(t)}, \quad y(t) = \frac{Y(t)}{W(t)}, \quad z(t) = \frac{Z(t)}{W(t)}.$$

Although this appears to introduce an unnecessary complication, the division by a second polynomial brings certain advantages:

- They describe perfect circles, ellipses, parabolas and hyperbolas, whereas non-rational curves can only approximate these curves.
- They are invariant of their control points when subjected to rotation, scaling, translation and perspective transformations, whereas non-rational curves lose this geometric integrity.
- They allow weights to be used at the control points to push and pull the curve.

© Springer Nature Switzerland AG 2019
J. Vince, *Calculus for Computer Graphics*,
https://doi.org/10.1007/978-3-030-11376-6_14

Fig. 14.1 The construction
of a uniform non-rational
B-spline curve

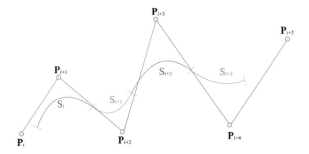

An explanation of uniform and non-uniform types is best left until you understand the idea of splines. So, without knowing the meaning of uniform, let's begin with uniform B-splines.

14.2.1 Uniform B-Splines

A B-spline is constructed from a string of curve segments whose geometry is determined by a group of local control points. These curves are known as *piecewise polynomials*. A curve segment does not have to pass through a control point, although this may be desirable at the two end points.

Cubic B-splines are very common, as they provide a geometry that is one step away from simple quadratics, and possess continuity characteristics that make the joins between the segments invisible. In order to understand their construction consider the scenario in Fig. 14.1. Here we see a group of $(m + 1)$ control points $\mathbf{P}_0$, $\mathbf{P}_1$, $\mathbf{P}_2, \ldots, \mathbf{P}_m$ which determine the shape of a cubic curve constructed from a series of curve segments $\mathbf{S}_0, \mathbf{S}_1, \mathbf{S}_2, \ldots, \mathbf{S}_{m-3}$.

As the curve is cubic, curve segment $\mathbf{S}_i$ is influenced by $\mathbf{P}_i, \mathbf{P}_{i+1}, \mathbf{P}_{i+2}, \mathbf{P}_{i+3}$, and curve segment $\mathbf{S}_{i+1}$ is influenced by $\mathbf{P}_{i+1}, \mathbf{P}_{i+2}, \mathbf{P}_{i+3}, \mathbf{P}_{i+4}$. And as there are $(m + 1)$ control points, there are $(m - 2)$ curve segments.

A single segment $\mathbf{S}_i(t)$ of a B-spline curve is defined by

$$\mathbf{S}_i(t) = \sum_{r=0}^{3} \mathbf{P}_{i+r} B_r(t), \quad \text{for } 0 \le t \le 1$$

where

$$B_0(t) = \tfrac{1}{6}\left(-t^3 + 3t^2 - 3t + 1\right) = \tfrac{1}{6}(1 - t)^3 \tag{14.1}$$
$$B_1(t) = \tfrac{1}{6}\left(3t^3 - 6t^2 + 4\right) \tag{14.2}$$
$$B_2(t) = \tfrac{1}{6}\left(-3t^3 + 3t^2 + 3t + 1\right) \tag{14.3}$$
$$B_3(t) = \tfrac{1}{6}t^3. \tag{14.4}$$

Fig. 14.2 The B-spline basis functions

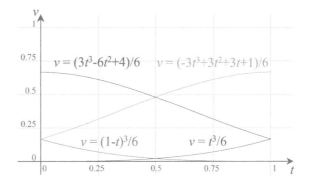

These are the B-spline *basis functions* and are shown in Fig. 14.2.

Although it is not apparent, these four curve segments are part of one curve. The basis function $B_3(t)$ starts at zero and rises to ≈ 0.166 at $t = 1$. It is taken over by $B_2(t)$ at $t = 0$, which rises to ≈ 0.166 at $t = 1$. The next segment is $B_1(t)$ and takes over at $t = 0$ and falls to ≈ 0.166 at $t = 1$. Finally, $B_0(t)$ takes over at ≈ 0.166 and falls to zero at $t = 1$. Equations (14.1)–(14.4) are represented in matrix form by

$$Q_1(t) = [t^3 \ t^2 \ t \ 1] \frac{1}{6} \begin{bmatrix} -1 & 3 & -3 & 1 \\ 3 & -6 & 3 & 0 \\ -3 & 0 & 3 & 0 \\ 1 & 4 & 1 & 0 \end{bmatrix} \begin{bmatrix} P_i \\ P_{i+1} \\ P_{i+2} \\ P_{i+3} \end{bmatrix}. \tag{14.5}$$

Let's now illustrate how (14.5) works. We first identify the control points P_i, P_{i+1}, P_{i+2}, etc. Let these be (0, 1), (1, 3), (2, 0), (4, 1), (4, 3), (2, 2) and (2, 3). They can be seen in Fig. 14.3 connected together by straight lines. If we take the first four control points: (0, 1), (1, 3), (2, 0), (4, 1), and subject the x- and y-coordinates to the matrix in (14.5) over the range $0 \le t \le 1$ we obtain the first B-spline curve segment shown in Fig. 14.3. If we move along one control point and take the next group of control points (1, 3), (2, 0), (4, 1), (4, 3), we obtain the second B-spline curve segment. This is repeated a further two times.

Figure 14.3 shows the four curve segments, and it is obvious that even though there are four discrete segments, they join together perfectly. This is no accident. The slopes at the end points of the basis curves are designed to match the slopes of their neighbours and ultimately keep the geometric curve continuous.

14.2.2 B-Spline Continuity

Constructing curves from several segments can only succeed if the slope of the abutting curves match. As we are dealing with curves whose slopes are changing everywhere, it will be necessary to ensure that even the rate of change of slopes is

Fig. 14.3 Four curve segments forming a B-spline curve

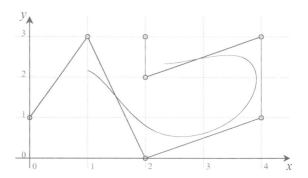

matched at the join. This aspect of curve design is called *geometric continuity* and is determined by the continuity properties of the basis function. Let's explore such features.

The *first level* of curve continuity C^0, ensures that the physical end of one basis curve corresponds with the following, e.g. $\mathbf{S}_i(1) = \mathbf{S}_{i+1}(0)$. We know that this occurs from the basis graphs shown in Fig. 14.2. The *second level* of curve continuity C^1, ensures that the slope at the end of one basis curve matches that of the following curve. This is confirmed by differentiating the basis functions (14.1)–(14.4):

$$B_0'(t) = \tfrac{1}{6}\left(-3t^2 + 6t - 3\right) \tag{14.6}$$
$$B_1'(t) = \tfrac{1}{6}\left(9t^2 - 12t\right) \tag{14.7}$$
$$B_2'(t) = \tfrac{1}{6}\left(-9t^2 + 6t + 3\right) \tag{14.8}$$
$$B_3'(t) = \tfrac{1}{6}\left(3t^2\right). \tag{14.9}$$

Evaluating (14.6)–(14.9) for $t = 0$ and $t = 1$, we discover the slopes 0.5, 0, -0.5, 0 for the joins between B_3, B_2, B_1, B_0. The *third level* of curve continuity C^2, ensures that the rate of change of slope at the end of one basis curve matches that of the following curve. This is confirmed by differentiating (14.6)–(14.9):

$$B_0''(t) = -t + 1 \tag{14.10}$$
$$B_1''(t) = 3t - 2 \tag{14.11}$$
$$B_2''(t) = -3t + 1 \tag{14.12}$$
$$B_3''(t) = t. \tag{14.13}$$

Evaluating (14.10)–(14.13) for $t = 0$ and $t = 1$, we discover the values 1, 2, 1, 0 for the joins between B_3, B_2, B_1, B_0. These combined continuity results are tabulated in Table 14.1.

Table 14.1 Continuity properties of cubic B-splines

t			t			t		
C^0	0	1	C^1	0	1	C^2	0	1
$B_3(t)$	0	1/6	$B_3'(t)$	0	0.5	$B_3''(t)$	0	1
$B_2(t)$	1/6	2/3	$B_2'(t)$	0.5	0	$B_2''(t)$	1	-2
$B_1(t)$	2/3	1/6	$B_1'(t)$	0	-0.5	$B_1''(t)$	-2	1
$B_0(t)$	1/6	0	$B_0'(t)$	-0.5	0	$B_0''(t)$	1	0

14.3 Derivatives of a Bézier Curve

In this section I describe how to calculate the first derivative of a Bézier curve by differentiating the basis function.

A Bézier curve with $n + 1$ control points $\mathbf{P}_0$, $\mathbf{P}_1, \ldots, \mathbf{P}_n$ is defined by

$$\mathbf{r}(t) = \sum_{i=0}^{n} B_{n,i}(t)\mathbf{P}_i$$

where the basis function is

$$B_{n,i}(t) = \frac{n!}{i!(n-i)!}t^i(1-t)^{n-i}.$$

For example, a quadratic Bézier curve with 3 control points, $n = 2$:

$$B_{2,i}(t) = \frac{2!}{i!(2-i)!}t^i(1-t)^{2-i}$$

$$B_{2,0}(t) = \frac{2!}{0!2!}t^0(1-t)^{2-0} \qquad = (1-t)^2$$

$$B_{2,1}(t) = \frac{2!}{1!(2-1)!}t^1(1-t)^{2-1} \qquad = 2t(1-t)$$

$$B_{2,2}(t) = \frac{2!}{2!(2-2)!}t^2(1-t)^{2-2} \qquad = t^2$$

and a cubic Bézier curve with 4 control points, $n = 3$:

$$B_{3,i}(t) = \frac{3!}{i!(3-i)!}t^i(1-t)^{3-i}$$

$$B_{3,0}(t) = \frac{3!}{0!3!}t^0(1-t)^{3-0} \qquad = (1-t)^3$$

$$B_{3,1}(t) = \frac{3!}{1!(3-1)!}t^1(1-t)^{3-1} \qquad = 3t(1-t)^2$$

$$B_{3,2}(t) = \frac{3!}{2!(3-2)!} t^2 (1-t)^{3-2} \quad = 3t^2(1-t)$$

$$B_{3,3}(t) = \frac{3!}{2!(3-3)!} t^2 (1-t)^{3-3} \quad = t^3.$$

Therefore, for a 3D curve with 4 control points $P_i(x_i, y_i, z_i)$, $0 \le i \le 3$:

$$\mathbf{r}(t) = \begin{bmatrix} x(t) \\ y(t) \\ z(t) \end{bmatrix}$$

$$x(t) = (1-t)^3 x_0 + 3t(1-t)^2 x_1 + 3t^2(1-t) x_2 + t^3 x_3$$
$$y(t) = (1-t)^3 y_0 + 3t(1-t)^2 y_1 + 3t^2(1-t) y_2 + t^3 y_3$$
$$z(t) = (1-t)^3 z_0 + 3t(1-t)^2 z_1 + 3t^2(1-t) z_2 + t^3 z_3.$$

To find the derivative of $\mathbf{r}(t)$, we first differentiate the basis function $B_{3,i}(t)$:

$$B'_{3,0}(t) = -3(1-t)^2 \qquad\qquad = -3B_{2,0}(t)$$
$$B'_{3,1}(t) = 3(1-t)^2 - 6t(1-t) \qquad = 3B_{2,0}(t) - 3B_{2,1}(t)$$
$$B'_{3,2}(t) = -3t^2 + 6t(1-t) \qquad\quad = 3B_{2,1}(t) - 3B_{2,2}(t)$$
$$B'_{3,3}(t) = 3t^2 \qquad\qquad\qquad\quad = 3B_{2,2}(t)$$

where we see that the derivative of a cubic Bézier curve is expressed in terms of a quadratic Bézier curve; consequently:

$$\mathbf{r}'(t) = -3\mathbf{P}_0 B_{2,0}(t) + 3\mathbf{P}_1 \left[B_{2,0}(t) - B_{2,1}(t) \right] + 3\mathbf{P}_2 \left[B_{2,1}(t) - B_{2,2}(t) \right] + 3\mathbf{P}_3 B_{2,2}(t)$$
$$= 3(\mathbf{P}_1 - \mathbf{P}_0) B_{2,0}(t) + 3(\mathbf{P}_2 - \mathbf{P}_1) B_{2,1}(t) + 3(\mathbf{P}_3 - \mathbf{P}_2) B_{2,2}(t). \qquad (14.14)$$

In order to generalise (14.14) we differentiate the basis function:

$$\frac{d}{dt} B_{n,i} = -(n-i) \frac{n!}{i!(n-i)!} t^i (1-t)^{n-i-1} + i \frac{n!}{i!(n-i)!} t^{i-1} (1-t)^{n-i}$$

$$= -(n-i) \frac{n(n-1)!}{i!(n-i)(n-1-i)!} t^i (1-t)^{n-1-i} + i \frac{n(n-1)!}{i(i-1)!(n-i)!} t^{i-1} (1-t)^{n-i}$$

$$= -n \frac{(n-1)!}{i!(n-1-i)!} t^i (1-t)^{n-1-i} + n \frac{(n-1)!}{(i-1)!(n-i)!} t^{i-1} (1-t)^{n-i}$$

$$B'_{n,i} = -n B_{n-1,i}(t) + n B_{n-1,i-1}(t)$$
$$= n \left(B_{n-1,i-1}(t) - B_{n-1,i}(t) \right).$$

Therefore,

$$\mathbf{r}'(t) = \sum_{i=0}^{n} \mathbf{P}_i B'_{n,i}(t)$$

$$= n \sum_{i=0}^{n} \mathbf{P}_i \left(B_{n-1,i-1}(t) - B_{n-1,i}(t)\right)$$

$$= n \sum_{i=0}^{n} \mathbf{P}_i B_{n-1,i-1}(t) - n \sum_{i=0}^{n} \mathbf{P}_i B_{n-1,i}(t).$$

In the second sum

$$n \sum_{i=0}^{n} \mathbf{P}_i B_{n-1,i}(t)$$

when $i = n$, $B_{n-1,n} = 0$, which permits the range of i to be reduced to $n - 1$:

$$\mathbf{r}'(t) = n \sum_{i=0}^{n} \mathbf{P}_i B_{n-1,i-1}(t) - n \sum_{i=0}^{n-1} \mathbf{P}_i B_{n-1,i}(t).$$

Next, the first sum is adjusted to sum to $n - 1$:

$$\mathbf{r}'(t) = n \sum_{i=0}^{n-1} \mathbf{P}_{i+1} B_{n-1,i}(t) - n \sum_{i=0}^{n-1} \mathbf{P}_i B_{n-1,i}(t)$$

$$\mathbf{r}'(t) = n \sum_{i=0}^{n-1} (\mathbf{P}_{i+1} - \mathbf{P}_i) B_{n-1,i}(t).$$

For a cubic, $n = 3$:

$$\mathbf{r}'(t) = 3 \, (\mathbf{P}_1 - \mathbf{P}_0) \, B_{2,0}(t) + 3 \, (\mathbf{P}_2 - \mathbf{P}_1) \, B_{2,1}(t) + 3 \, (\mathbf{P}_3 - \mathbf{P}_2) \, B_{2,2}(t)$$

which is the same as (14.14).

Let's calculate the first derivative of the following cubic Bézier curve.

The control points are $P_0 = (0, 0)$, $P_1 = (0, 1)$, $P_2 = (1, 1)$, $P_3 = (2, 0)$ as shown in Fig. 14.4. Therefore,

$$
\begin{aligned}
x'(t) &= 3(0 - 0)(1 - t)^2 + 3(1 - 0)2t(1 - t) + 3(2 - 1)t^2 \\
&= 6t(1 - t) + 3t^2 \\
&= 6t - 3t^2 \\
y'(t) &= 3(1 - 0)(1 - t)^2 + 3(1 - 1)2t(1 - t) + 3(0 - 1)t^2 \\
&= 3(1 - t)^2 - 3t^2 \\
&= 3 - 6t.
\end{aligned}
$$

The original parametric functions must give the same result:

$$x(t) = (1-t)^3 0 + 3t(1-t)^2 0 + 3t^2(1-t)1 + t^3 2$$
$$= 3t^2 - t^3$$
$$x'(t) = 6t - 3t^2$$
$$y(t) = (1-t)^3 0 + 3t(1-t)^2 1 + 3t^2(1-t)1 + t^3 0$$
$$= 3t(1 - 2t + t^2) + 3t^2 - 3t^3$$
$$= 3t - 3t^2$$
$$y'(t) = 3 - 6t.$$

The derivatives at $t = 0$ and $t = 1$ are

$$x'(0) = 0$$
$$y'(0) = 3$$
$$x'(1) = 3$$
$$y'(1) = -3.$$

The derivatives $x'(t)$ and $y'(t)$ are with respect to t. To find dy/dx we divide $y'(t)$ by $x'(t)$:

$$\frac{dy}{dx} = \frac{\frac{dy}{dt}}{\frac{dx}{dt}} = \frac{3 - 6t}{6t - 3t^2} = \frac{1 - 2t}{2t - t^2}.$$

When $t = 0$, $\frac{dy}{dx} = \infty$, and when $t = 1$, $\frac{dy}{dx} = -1$, which correspond to the slopes of the first and last line segments respectively. See Fig. 14.4. This is always the case, because:

$$\mathbf{r}'(t) = 3(\mathbf{P}_1 - \mathbf{P}_0) B_{2,0}(t) + 3(\mathbf{P}_2 - \mathbf{P}_1) B_{2,1}(t) + 3(\mathbf{P}_3 - \mathbf{P}_2) B_{2,2}(t)$$
$$\mathbf{r}'(0) = 3(\mathbf{P}_1 - \mathbf{P}_0) B_{2,0}(t)$$
$$x'(0) = 3(x_1 - x_0) B_{2,0}(t)$$
$$y'(0) = 3(y_1 - y_0) B_{2,0}(t)$$
$$\frac{dy}{dx} = \frac{y'(0)}{x'(0)} = \frac{y_1 - y_0}{x_1 - x_0}$$
$$\mathbf{r}'(1) = 3(\mathbf{P}_3 - \mathbf{P}_2) B_{2,2}(t)$$
$$x'(1) = 3(x_3 - x_2) B_{2,2}(t)$$
$$y'(1) = 3(y_3 - y_2) B_{2,2}(t)$$
$$\frac{dy}{dx} = \frac{y'(1)}{x'(1)} = \frac{y_3 - y_2}{x_3 - x_2}.$$

Fig. 14.4 A 2D cubic Bézier curve

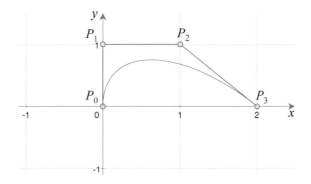

The second derivative is stated without proof as

$$\mathbf{r}''(t) = \sum_{i=0}^{n-2} n(n-1)\,(\mathbf{P}_{i+2} - 2\mathbf{P}_{i+1} + \mathbf{P}_i)\,B_{n-2,i}(t).$$

Using the example shown in Fig. 14.4, $P_0 = (0,0)$, $P_1 = (0,1)$, $P_2 = (1,1)$, $P_3 = (2,0)$

$$x''(t) = 3 \times 2\,[(1 - 2 \times 0 + 0)\,(1 - t) + (2 - 2 \times 1 + 0)\,t]$$
$$= 6(1 - t)$$
$$y''(t) = 3 \times 2\,[(1 - 2 \times 1 + 0)\,(1 - t) + (0 - 2 \times 1 + 1)\,t]$$
$$= 6\,[-(1 - t) - t] = -6.$$

The first derivative equals

$$\frac{dy}{dx} = \frac{1 - 2t}{2t - t^2}$$

which equals zero, when $t = 0.5$. The second derivative equals

$$\frac{d^2 y}{dx^2} = \frac{y''(t)}{x''(2)} = \frac{-6}{6(1 - t)}$$

which is negative at $t = 0.5$, therefore, there is a maximum value at this point.

14.4 Summary

Fortunately, geometric continuity is reasonably easy to illustrate: it's just a question of differentiating the basis functions. However, there are many other types of curves, where the same technique can be applied.

Chapter 15
Curvature

15.1 Introduction

In this chapter I describe the mathematical definition of curvature, and show how to compute the curvature of a circle, helix, parabola, sine curve, Bézier curve, and a graph described by an explicit equation.

15.2 Curvature

When we hold a curved object, we can tell immediately the tightness of the curved surface. Similarly, when driving along a twisting roadway, the forces on our body reflect the curvature of the path taken by the vehicle. Curvature is expressed mathematically in a variety of ways, and we will see the benefits and drawbacks of each one.

With reference to Fig. 15.1, the curve at point P is approximately equal to part of a circular arc with radius R. Therefore, the curvature κ (kappa), of the curve at P is defined as $\kappa = 1/R$. The reciprocal of R is chosen so that a small radius corresponds to a large curvature, and a large radius, a small curvature. One can see that the curve at point Q is almost a straight line, which corresponds to a very large circle, and therefore a small curvature. This circle is called the *osculating circle*.

We can see that some curves such as a circular arc, and a linear helix, have a constant curvature, whereas a parabola, elliptical arc, quadratic curve, etc., have different degrees of curvature along their length.

In order to calculate κ, we investigate how fast unit tangent vectors change along the curve. Unit vectors are chosen, otherwise the tangent-vector length influences the rate of change. Figure 15.2 shows a curve with unit tangent vectors placed at the points A, B, C and D. It is clear that at points of high curvature, the associated unit tangent vectors change faster than those at points of low curvature. This measure of curvature is expressed as

© Springer Nature Switzerland AG 2019
J. Vince, *Calculus for Computer Graphics*,
https://doi.org/10.1007/978-3-030-11376-6_15

Fig. 15.1 The curvature at
P is defined by $\kappa = \frac{1}{R}$

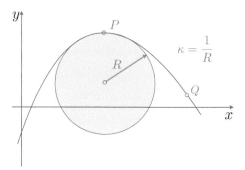

Fig. 15.2 The unit tangent
vectors at different points
along a curve

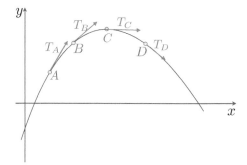

$$\kappa = \left\|\frac{d\mathbf{T}}{ds}\right\|.$$

Note that the derivative is relative to the arc length s, which can be a problem to compute, and the reason for taking the absolute value is to remove any negative sign that may arise. Curvature is regarded as an unsigned quantity. Let's see how this definition behaves in practice.

15.2.1   Curvature of a Circle

Consider the vector-valued function $\mathbf{r}(t)$ for a circle of radius r:

$$\mathbf{r}(t) = \begin{bmatrix} r\cos t \\ r\sin t \end{bmatrix}, \quad t \in [0, 2\pi]$$

its derivative is the tangent vector on the curve:

$$\mathbf{r}'(t) = \begin{bmatrix} -r\sin t \\ r\cos t \end{bmatrix}.$$

The unit tangent vector function $\mathbf{T}(t)$ is

$$\mathbf{T}(t) = \frac{\mathbf{r}'(t)}{||\mathbf{r}'(t)||}$$

but

$$||\mathbf{r}'(t)|| = \sqrt{(-r\sin t)^2 + (r\cos t)^2} = r$$

therefore,

$$\mathbf{T}(t) = \frac{\mathbf{r}'(t)}{r} = \begin{bmatrix} -\sin t \\ \cos t \end{bmatrix}.$$

Now,

$$\kappa = \left\|\frac{d\mathbf{T}}{ds}\right\|$$

but as we don't know $d\mathbf{T}/ds$, we use the chain rule to redefine κ:

$$\kappa = \frac{\left\|\dfrac{d\mathbf{T}}{dt}\right\|}{\left\|\dfrac{ds}{dt}\right\|}.$$

We have already seen that

$$\frac{ds}{dt} = ||\mathbf{r}'(t)||$$

which equals r, and

$$\frac{d\mathbf{T}}{dt} = \begin{bmatrix} -\cos t \\ -\sin t \end{bmatrix}$$

$$\left\|\frac{d\mathbf{T}}{dt}\right\| = \sqrt{(-\cos t)^2 + (-\sin t)^2} = 1$$

therefore,

$$\kappa = \frac{1}{r}$$

which agrees with the original definition of curvature.

15.2.2 Curvature of a Helix

For this example we employ a helix with a constant pitch and radius a:

$$\mathbf{r}(t) = \begin{bmatrix} a\cos t \\ a\sin t \\ bt \end{bmatrix}, \quad t \in [0, \ 2\pi]$$

its derivative is the tangent vector on the curve:

$$\mathbf{r}'(t) = \begin{bmatrix} -a\sin t \\ a\cos t \\ b \end{bmatrix}.$$

The unit tangent vector function $\mathbf{T}(t)$ is

$$\mathbf{T}(t) = \frac{\mathbf{r}'(t)}{||\mathbf{r}'(t)||}$$

where

$$\begin{aligned} ||\mathbf{r}'(t)|| &= \sqrt{(-a\sin t)^2 + (a\cos t)^2 + b^2} \\ &= \sqrt{a^2\sin^2 t + a^2\cos^2 t + b^2} \\ &= \sqrt{a^2\left(\sin^2 t + \cos^2 t\right) + b^2} \\ &= \sqrt{a^2 + b^2} \end{aligned}$$

therefore,

$$\mathbf{T}(t) = \frac{\mathbf{r}'(t)}{\sqrt{a^2 + b^2}} = \frac{1}{\sqrt{a^2 + b^2}} \begin{bmatrix} -a\sin t \\ a\cos t \\ b \end{bmatrix}.$$

Now,

$$\kappa = \frac{\left|\left|\dfrac{d\mathbf{T}}{dt}\right|\right|}{\left|\left|\dfrac{ds}{dt}\right|\right|}.$$

where

$$\frac{d\mathbf{T}}{dt} = \frac{1}{\sqrt{a^2 + b^2}} \begin{bmatrix} -a\cos t \\ -a\sin t \\ 0 \end{bmatrix}$$

$$\left|\left|\frac{d\mathbf{T}}{dt}\right|\right| = \frac{a}{\sqrt{a^2 + b^2}}$$

$$\frac{ds}{dt} = ||\mathbf{r}'(t)|| = \sqrt{a^2 + b^2}$$

Fig. 15.3 A helix where $x = 3\cos t$, $y = 3\sin t$, $z = 0.25t$ and $t \in [0, 4\pi]$

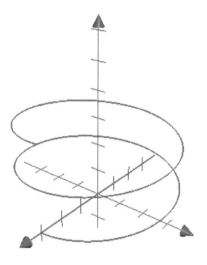

$$\kappa = \frac{a}{a^2 + b^2}.$$

Note that when $b = 0$, $\kappa = 1/a$, which is what one would expect.

Let's illustrate this with an example where $a = 3$ and $b = 0.25$, as shown in Fig. 15.3. Therefore,

$$\kappa = \frac{3}{3^2 + 0.25^2} = \frac{3}{9.0625} \approx 0.331.$$

Thus, the curvature is ≈ 0.331, and the radius of the osculating circle is ≈ 3.0208, which is slightly larger than the underling circle for the helix.

15.2.3 Curvature of a Parabola

A simple parabola such as $y = x^2$, is written as vector-valued function as

$$\mathbf{r}(t) = \begin{bmatrix} t \\ t^2 \end{bmatrix}$$

its derivative is the tangent vector on the curve:

$$\mathbf{r}'(t) = \begin{bmatrix} 1 \\ 2t \end{bmatrix}.$$

The unit tangent vector function $\mathbf{T}(t)$ is

$$\mathbf{T}(t) = \frac{\mathbf{r}'(t)}{||\mathbf{r}'(t)||}$$

where

$$||\mathbf{r}'(t)|| = \sqrt{1 + 4t^2}$$

therefore,

$$\mathbf{T}(t) = \frac{\mathbf{r}'(t)}{\sqrt{1 + 4t^2}} = \begin{bmatrix} 1/\sqrt{1 + 4t^2} \\ 2t/\sqrt{1 + 4t^2} \end{bmatrix}.$$

Differentiating $y = 1/\left(1 + 4t^2\right)^{\frac{1}{2}}$:

Let $u = 1 + 4t^2$:

$$y = u^{-\frac{1}{2}}$$

$$\frac{dy}{du} = -\frac{1}{2}u^{-\frac{3}{2}} = -\frac{1}{2(1 + 4t^2)^{\frac{3}{2}}}$$

$$\frac{du}{dt} = 8t$$

$$\frac{dy}{dt} = \frac{dy}{du} \cdot \frac{du}{dt} = -\frac{8t}{2(1 + 4t^2)^{\frac{3}{2}}}$$

$$= -\frac{4t}{(1 + 4t^2)^{\frac{3}{2}}}$$

Differentiating the quotient $y = 2t/(1 + 4t^2)^{\frac{1}{2}}$:

$$\frac{dy}{dt} = \frac{2(1 + 4t^2)^{\frac{1}{2}} - 2t\frac{d}{dx}(1 + 4t^2)^{\frac{1}{2}}}{1 + 4t^2}$$

$$= \frac{2(1 + 4t^2)^{\frac{1}{2}} - 2t\left(\frac{1}{2}(1 + 4t^2)^{-\frac{1}{2}}(8t)\right)}{1 + 4t^2}$$

$$= \frac{2(1 + 4t^2)^{\frac{1}{2}} - 8t^2(1 + 4t^2)^{-\frac{1}{2}}}{1 + 4t^2}$$

$$= \frac{2(1 + 4t^2) - 8t^2}{(1 + 4t^2)^{\frac{3}{2}}}$$

$$= \frac{2}{(1 + 4t^2)^{\frac{3}{2}}}$$

therefore,

$$\frac{d\mathbf{T}}{dt} = \begin{bmatrix} -\frac{4t}{(1 + 4t^2)^{\frac{3}{2}}} \\ \frac{2}{(1 + 4t^2)^{\frac{3}{2}}} \end{bmatrix}$$

Fig. 15.4 The parabola $y = x^2$ with the osculating circle, radius 0.5

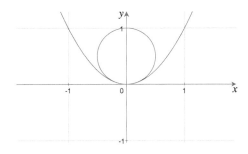

and

$$\left\| \frac{d\mathbf{T}}{dt} \right\| = \sqrt{\left(-\frac{4t}{(1 + 4t^2)^{\frac{3}{2}}} \right)^2 + \left(\frac{2}{(1 + 4t^2)^{\frac{3}{2}}} \right)^2}$$

$$= \sqrt{\frac{16t^2 + 4}{(1 + 4t^2)^3}} = \sqrt{\frac{4(1 + 4t^2)}{(1 + 4t^2)^3}}$$

$$= \frac{2}{1 + 4t^2}.$$

But

$$\kappa = \frac{\left\| \dfrac{d\mathbf{T}}{dt} \right\|}{\left\| \dfrac{ds}{dt} \right\|}.$$

where

$$\frac{ds}{dt} = (1 + 4t^2)^{\frac{1}{2}}$$

therefore,

$$\kappa(t) = \frac{2}{(1 + 4t^2)(1 + 4t^2)^{\frac{1}{2}}} = \frac{2}{(1 + 4t^2)^{\frac{3}{2}}}.$$

When $t = 0$, then $\kappa = 2$, as shown in Fig. 15.4. Naturally, as one moves away from this trough in the curve, the radius of curvature increases, and κ decreases.

I am sure you will agree, that the above proof is incredibly long, and there must be a better way. Fortunately, there is, and it is shown in the next section.

15.2.4 Parametric Plane Curve

Given a 2D parametric plane curve:

$$\mathbf{r}(t) = \begin{bmatrix} x(t) \\ y(t) \end{bmatrix}$$

its curvature is

$$\kappa(t) = \frac{||x'(t) \cdot y''(t) - y'(t) \cdot x''(t)||}{\left(x'(t)^2 + y'(t)^2\right)^{\frac{3}{2}}}$$

where

$$x' = \frac{dx}{dt}, \qquad y' = \frac{dy}{dt}$$
$$x'' = \frac{d^2x}{dt^2}, \qquad y'' = \frac{d^2y}{dt^2}.$$

Let's try a parabola

$$x(t) = t, \qquad y(t) = t^2$$
$$x'(t) = 1, \qquad y'(t) = 2t$$
$$x''(t) = 0, \qquad y''(t) = 2.$$

Therefore,

$$\kappa = \frac{||1 \times 2 - 2t \times 0||}{\left(1 + 4t^2\right)^{\frac{3}{2}}}$$
$$= \frac{2}{\left(1 + 4t^2\right)^{\frac{3}{2}}}.$$

When $t = 0$, $\kappa = 2$. This is a much simpler method of calculating curvature, and there is a 3D version which we examine later.

Now let's try a sine curve:

$$x(t) = t, \qquad y(t) = \sin t$$
$$x'(t) = 1, \qquad y'(t) = \cos t$$
$$x''(t) = 0, \qquad y''(t) = -\sin t.$$

Therefore,

$$\kappa = \frac{|-\sin t|}{\left(1 + \cos^2 t\right)^{\frac{3}{2}}}.$$

When $t = 0$, $\kappa = 0$, and when $t = \pi/2$, $\kappa = 1$, as shown in Fig. 15.5.

In order to calculate the curvature of 3D parametric curves, we use (15.1):

Fig. 15.5 A sine curve
$y = \sin x$ with the osculating
circle, radius 1

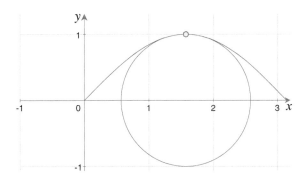

$$\kappa = \frac{\sqrt{(z'' \cdot y' - y'' \cdot z')^2 + (x'' \cdot z' - z'' \cdot x')^2 + (y'' \cdot x' - x'' \cdot y')^2}}{\left(x'^2 + y'^2 + z'^2\right)^{\frac{3}{2}}}. \quad (15.1)$$

Given a vector-valued function:

$$\mathbf{r}(t) = \begin{bmatrix} x(t) \\ y(t) \\ z(t) \end{bmatrix}, \quad \mathbf{r}'(t) = \begin{bmatrix} x'(t) \\ y'(t) \\ z'(t) \end{bmatrix}, \quad \mathbf{r}''(t) = \begin{bmatrix} x''(t) \\ y''(t) \\ z''(t) \end{bmatrix}.$$

$\mathbf{r}'(t)$ is the tangent vector to the curve, and $\mathbf{r}''(t)$ is the rate of change of the tangent vector. The cross-product $\mathbf{r}'(t) \times \mathbf{r}''(t)$ is a measure of the curvature, whose magnitude is the numerator in (15.1). The denominator $\left(x'^2 + y'^2 + z'^2\right)^{\frac{3}{2}}$, is the required scaling factor. Let's use (15.1) to calculate the curvature of a constant pitch helix.

$$\mathbf{r}(t) = \begin{bmatrix} a\cos t \\ a\sin t \\ bt \end{bmatrix}, \quad \mathbf{r}'(t) = \begin{bmatrix} -a\sin t \\ a\cos t \\ b \end{bmatrix}, \quad \mathbf{r}''(t) = \begin{bmatrix} -a\cos t \\ -a\sin t \\ 0 \end{bmatrix}.$$

Therefore,

$$\begin{aligned}
\kappa &= \frac{\sqrt{(-ab\sin t)^2 + (-ab\cos t)^2 + \left(a^2\sin^2 t + a^2\cos^2 t\right)^2}}{\left(a^2\cos^2 t + a^2\sin^2 t + b^2 t^2\right)^{\frac{3}{2}}} \\
&= \frac{\sqrt{a^2 b^2 \sin^2 t + a^2 b^2 \cos^2 t + a^4}}{\left(a^2 + b^2\right)^{\frac{3}{2}}} \\
&= \frac{\sqrt{a^2 b^2 + a^4}}{\left(a^2 + b^2\right)^{\frac{3}{2}}} \\
&= \frac{a\left(a^2 + b^2\right)^{\frac{1}{2}}}{\left(a^2 + b^2\right)^{\frac{3}{2}}}
\end{aligned}$$

$$= \frac{a}{a^2 + b^2}$$

which agrees with the result for the previous helix.

15.2.5   Curvature of a Graph

When a curve is expressed as an explicit function, the curvature κ is

$$\kappa = \frac{\left|y''(t)\right|}{\left(1 + y'(t)^2\right)^{\frac{3}{2}}}.$$

For a sine curve:

$$y = \sin t$$
$$y'(t) = \cos t$$
$$y''(t) = -\sin t.$$

Therefore,

$$\kappa = \frac{\left|-\sin t\right|}{\left(1 + \cos^2 t\right)^{\frac{3}{2}}}$$

which is the same as a parametric plane curve.
   Applying this formula for a parabola:

$$y = t^2$$
$$y'(t) = 2t$$
$$y''(t) = 2$$

$$\kappa = \frac{2}{\left(1 + 4t^2\right)^{\frac{3}{2}}}$$

which agrees with the previous result, and is much simpler.

15.2.6   Curvature of a 2D Quadratic Bézier Curve

A 2D quadratic Bézier curve is defined as

$$\mathbf{r}(t) = \mathbf{P}_0(1 - t)^2 + 2\mathbf{P}_1 t(1 - t) + \mathbf{P}_2 t^2$$

Fig. 15.6 A Bézier curve with the osculating circle, radius 1

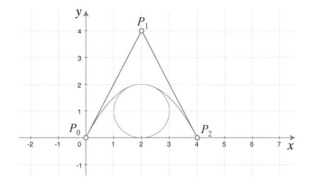

which has the following first and second derivatives:

$$\begin{aligned}
\mathbf{r}'(t) &= -2\mathbf{P}_0(1-t) + 2\mathbf{P}_1(1-2t) + 2\mathbf{P}_2 t \\
&= -2\mathbf{P}_0 + 2\mathbf{P}_0 t + 2\mathbf{P}_1 - 4\mathbf{P}_1 t + 2\mathbf{P}_2 t \\
&= 2(\mathbf{P}_1 - \mathbf{P}_0)(1-t) + 2(\mathbf{P}_2 - \mathbf{P}_1)t \\
&= 2[(\mathbf{P}_1 - \mathbf{P}_0)(1-t) + (\mathbf{P}_2 - \mathbf{P}_1)t] \\
\mathbf{r}''(t) &= 2(\mathbf{P}_0 - 2\mathbf{P}_1 + \mathbf{P}_2).
\end{aligned}$$

We can now use

$$\kappa(t) = \frac{\left\| x'(t) \cdot y''(t) - y'(t) \cdot x''(t) \right\|}{\left(x'(t)^2 + y'(t)^2 \right)^{\frac{3}{2}}} \tag{15.2}$$

to find the curvature. For example, Fig. 15.6 shows a Bézier curve with control points $P_0 = (0,0)$, $P_1 = (2,4)$, $P_2 = (4,0)$, which generate the following first and second derivatives at $t = 0.5$:

$$\begin{aligned}
x'(0.5) &= 2[2(1-0.5)+1] = 4 \\
y'(0.5) &= 2[4(1-0.5)-2] = 0 \\
x''(0.5) &= 2(0-4+4) = 0 \\
y''(0.5) &= 2(0-8+0) = -16.
\end{aligned}$$

Plugging these into (15.2), we get

$$\kappa(0.5) = \frac{\|-64 - 0\|}{(16+0)^{\frac{3}{2}}} = 1$$

which is confirmed by the unit-radius osculating circle in Fig. 15.6.

15.2.7   Curvature of a 2D Cubic Bézier Curve

A 2D cubic Bézier curve is defined as

$$\mathbf{r}(t) = \mathbf{P}_0(1-t)^3 + 3\mathbf{P}_1 t(1-t)^2 + 3\mathbf{P}_2 t^2(1-t) + \mathbf{P}_3 t^3$$

which has the following first and second derivatives:

$$
\begin{aligned}
\mathbf{r}'(t) &= -3\mathbf{P}_0(1-t)^2 + 3\mathbf{P}_1(1-4t+3t^2) + 3\mathbf{P}_2(2t-3t^2) + 3\mathbf{P}_3 t^2 \\
&= -3\mathbf{P}_0(1-t)^2 + 3\mathbf{P}_1(1-t)^2 - 3\mathbf{P}_1 2t(1-t) + 3\mathbf{P}_2 2t(1-t) - 3\mathbf{P}_2 t^2 + 3\mathbf{P}_3 t^2 \\
&= 3\left((\mathbf{P}_1 - \mathbf{P}_0)(1-t)^2 + (\mathbf{P}_2 - \mathbf{P}_1)2t(1-t) + (\mathbf{P}_3 - \mathbf{P}_2)t^2\right) \\
\mathbf{r}''(t) &= 3(\mathbf{P}_1 - \mathbf{P}_0)(-2+2t) + 3(\mathbf{P}_2 - \mathbf{P}_1)(2-4t) + 3(\mathbf{P}_3 - \mathbf{P}_2)2t \\
&= -3(\mathbf{P}_1 - \mathbf{P}_0)2(1-t) + 3(\mathbf{P}_2 - \mathbf{P}_1)2(1-t) - 3(\mathbf{P}_2 - \mathbf{P}_1)2t + 3(\mathbf{P}_3 - \mathbf{P}_2)2t \\
&= 6\left((\mathbf{P}_2 - 2\mathbf{P}_1 + \mathbf{P}_0)(1-t) + (\mathbf{P}_3 - 2\mathbf{P}_2 + \mathbf{P}_1)t\right)
\end{aligned}
$$

We can now use (15.2) to find the curvature. For example, Fig. 15.7 shows a Bézier curve with control points $\mathbf{P}_0 = (0,0)$, $\mathbf{P}_1 = (1,1)$, $\mathbf{P}_2 = (2,1)$, $P_3(3,0)$, which generate the following first and second derivatives at $t = 0.5$:

$$
\begin{aligned}
x'(0.5) &= 3\left(\tfrac{1}{4} + \tfrac{1}{2} + \tfrac{1}{4}\right) = 3 \\
y'(0.5) &= 3\left(\tfrac{1}{4} - \tfrac{1}{4}\right) = 0 \\
x''(0.5) &= 6\,(0) = 0 \\
y''(0.5) &= 6\left(-\tfrac{1}{2} - \tfrac{1}{2}\right) = -6.
\end{aligned}
$$

Plugging these into (15.2), we get

$$\kappa(0.5) = \frac{||-18||}{9^{\frac{3}{2}}} \approx 0.6667$$

which makes the radius of the osculating circle 1.5, as shown in Fig. 15.7.

15.3   Summary

Curvature has quite a simple definition, yet it some cases, requires tiresome levels of algebraic manipulation to secure an answer. Half the problem is choosing the most useful way of describing the original function.

Fig. 15.7 A cubic Bézier curve with the osculating circle, radius 1.5

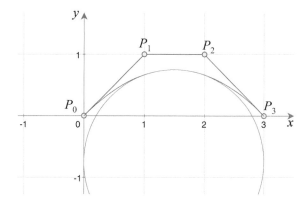

15.3.1 Summary of Formulae

Curvature κ

$$\kappa = \frac{1}{R}$$

where R is the radius of the osculating circle.

$$\kappa = \left\| \frac{d\mathbf{T}}{ds} \right\|.$$

where $\mathbf{T}$ is the unit tangent vector at a point along the arc length s.

$$\kappa = \frac{\left\| \dfrac{d\mathbf{T}}{dt} \right\|}{\left\| \dfrac{ds}{dt} \right\|}$$

where

$$\frac{ds}{dt} = \|\mathbf{r}'(t)\|.$$

Curvature of a Helix

$$\mathbf{r}(t) = \begin{bmatrix} a\cos t \\ a\sin t \\ bt \end{bmatrix}, \quad t \in [0, \, 2\pi]$$

$$\kappa = \frac{a}{a^2 + b^2}.$$

Curvature of a Parabola

$$\mathbf{r}(t) = \begin{bmatrix} t \\ t^2 \end{bmatrix}$$

$$\kappa(t) = \frac{2}{(1 + 4t^2)^{\frac{3}{2}}}.$$

Curvature of a 2D Parametric Plane Curve

$$\kappa(t) = \frac{\left\| x' \cdot y'' - y' \cdot x'' \right\|}{\left(x'^2 + y'^2 \right)^{\frac{3}{2}}}.$$

Curvature of a 3D Parametric Plane Curve

$$\kappa(t) = \frac{\sqrt{(z'' \cdot y' - y'' \cdot z')^2 + (x'' \cdot z' - z'' \cdot x')^2 + (y'' \cdot x' - x'' \cdot y')^2}}{\left(x'^2 + y'^2 + z'^2 \right)^{\frac{3}{2}}}.$$

Curvature of a Graph

$$\kappa(t) = \frac{\left\| y''(t) \right\|}{\left(1 + y'(t)^2 \right)^{\frac{3}{2}}}.$$

Chapter 16
Conclusion

Calculus is such a large subject, that everything one investigates leads to something else, and one is tempted to write about it and explain how and why it works. Consequently, when I started writing this book I had clear objectives about what to include and what to leave out. Having reached this final chapter, I feel that I have achieved this objective. There have been moments when I was tempted to include more topics and more examples and turn this book into similar books on Calculus that are extremely large and daunting to open.

Hopefully, the topics I have included will inspire you to read other books on Calculus and consolidate your knowledge and understanding of this important branch of mathematics.

© Springer Nature Switzerland AG 2019   291
J. Vince, *Calculus for Computer Graphics*,
https://doi.org/10.1007/978-3-030-11376-6_16

Appendix A
Limit of $(\sin \theta)/\theta$

This appendix proves that

$$\lim_{\theta \to 0} \frac{\sin \theta}{\theta} = 1, \quad \text{where } \theta \text{ is in radians.}$$

From high-school mathematics we know that $\sin \theta \approx \theta$, for small values of θ. For example:

$$\sin 0.1 \approx 0.099833$$
$$\sin 0.05 \approx 0.04998$$
$$\sin 0.01 \approx 0.0099998$$

and

$$\frac{\sin 0.1}{0.1} \approx 0.99833$$
$$\frac{\sin 0.05}{0.05} \approx 0.99958$$
$$\frac{\sin 0.01}{0.01} \approx 0.99998.$$

Therefore, we can reason that in the limit, as $\theta \to 0$:

$$\lim_{\theta \to 0} \frac{\sin \theta}{\theta} = 1.$$

Figure A.1 shows a graph of $(\sin \theta)/\theta$, which confirms this result. However, this is an observation, rather than a proof. So, let's pursue a geometric line of reasoning.

From Fig. A.2 we see as the circle's radius is unity, $OA = OB = 1$, and $AC = \tan \theta$. As part of the strategy, we need to calculate the area of the triangle $\triangle OAB$, the sector OAB and the $\triangle OAC$:

© Springer Nature Switzerland AG 2019
J. Vince, *Calculus for Computer Graphics*,
https://doi.org/10.1007/978-3-030-11376-6

Fig. A.1 Graph of $(\sin\theta)/\theta$

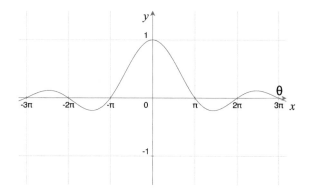

Fig. A.2 Unit radius circle
with trigonometric ratios

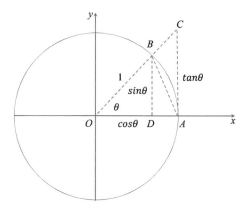

$$\text{Area of } \triangle OAB = \triangle ODB + \triangle DAB$$
$$= \tfrac{1}{2}\cos\theta \cdot \sin\theta + \tfrac{1}{2}(1 - \cos\theta) \cdot \sin\theta$$
$$= \tfrac{1}{2}\cos\theta \cdot \sin\theta + \tfrac{1}{2}\sin\theta - \tfrac{1}{2}\cos\theta \cdot \sin\theta$$
$$= \tfrac{1}{2}\sin\theta.$$
$$\text{Area of sector } OAB = \frac{\theta}{2\pi}\pi(1)^2 = \tfrac{1}{2}\theta.$$
$$\text{Area of } \triangle OAC = \tfrac{1}{2}(1)\tan\theta = \tfrac{1}{2}\tan\theta.$$

From the geometry of a circle, we know that

$$\tfrac{1}{2}\sin\theta < \tfrac{1}{2}\theta < \tfrac{1}{2}\tan\theta$$
$$\sin\theta < \theta < \frac{\sin\theta}{\cos\theta}$$
$$1 < \frac{\theta}{\sin\theta} < \frac{1}{\cos\theta}$$
$$1 > \frac{\sin\theta}{\theta} > \cos\theta$$

and as $\theta \to 0$, $\cos\theta \to 1$ and $\dfrac{\sin\theta}{\theta} \to 1$. This holds, even for negative values of θ, because

$$\frac{\sin(-\theta)}{-\theta} = \frac{-\sin\theta}{-\theta} = \frac{\sin\theta}{\theta}.$$

Therefore,

$$\lim_{\theta \to 0} \frac{\sin\theta}{\theta} = 1.$$

Appendix B
Integrating $\cos^n \theta$

This appendix shows how to evaluate $\int \cos^n \theta \, d\theta$.

We start with

$$\int \cos^n x \, dx = \int \cos x \cdot \cos^{n-1} x \, dx.$$

Let $u = \cos^{n-1} x$ and $v' = \cos x$, then

$$u' = -(n-1)\cos^{n-2} x \cdot \sin x$$

and

$$v = \sin x.$$

Integrating by parts:

$$\int uv' \, dx = uv - \int v u' \, dx + C$$

$$\int \cos^{n-1} x \cdot \cos x \, dx = \cos^{n-1} x \cdot \sin x + \int \sin x \cdot (n-1)\cos^{n-2} x \cdot \sin x \, dx + C$$

$$= \sin x \cdot \cos^{n-1} x + (n-1)\int \sin^2 x \cdot \cos^{n-2} x \, dx + C$$

$$= \sin x \cdot \cos^{n-1} x + (n-1)\int \left(1 - \cos^2 x\right) \cdot \cos^{n-2} x \, dx + C$$

$$= \sin x \cdot \cos^{n-1} x + (n-1)\int \cos^{n-2} \, dx - (n-1)\int \cos^n x \, dx + C$$

$$n \int \cos^n x \, dx = \sin x \cdot \cos^{n-1} x + (n-1)\int \cos^{n-2} \, dx + C$$

$$\int \cos^n x \, dx = \frac{\sin x \cdot \cos^{n-1} x}{n} + \frac{n-1}{n}\int \cos^{n-2} \, dx + C$$

where n is an integer, $\neq 0$.

© Springer Nature Switzerland AG 2019
J. Vince, *Calculus for Computer Graphics*,
https://doi.org/10.1007/978-3-030-11376-6

Similarly,

$$\int \sin^n x \, dx = -\frac{\cos x \cdot \sin^{n-1} x}{n} + \frac{n-1}{n} \int \sin^{n-2} \, dx + C.$$

For example,

$$\int \cos^3 x \, dx = \tfrac{1}{3} \sin x \cdot \cos^2 x + \tfrac{2}{3} \sin x + C.$$

Index

© Springer Nature Switzerland AG 2019
J. Vince, *Calculus for Computer Graphics*,
https://doi.org/10.1007/978-3-030-11376-6